SOLAR SILICON PROCESSES
Technologies, Challenges, and Opportunities

SOLAR SILICON PROCESSES

Technologies, Challenges, and Opportunities

edited by

Bruno Ceccaroli
Eivind Øvrelid
Sergio Pizzini

CRC Press
Taylor & Francis Group
Boca Raton London New York

CRC Press is an imprint of the
Taylor & Francis Group, an **informa** business

CRC Press
Taylor & Francis Group
6000 Broken Sound Parkway NW, Suite 300
Boca Raton, FL 33487-2742

First issued in paperback 2019

ISBN-13: 978-1-4987-4265-8 (hbk)
ISBN-13: 978-0-367-87478-0 (pbk)

Library of Congress Cataloging-in-Publication Data

Names: Ceccaroli, Bruno, editor. | Øvrelid, Eivind, editor. | Pizzini, Sergio, editor.
Title: Solar silicon processes : technologies, challenges, and opportunities / editors, Bruno Ceccaroli, Eivind Ovrelid, and Sergio Pizzini.
Description: Boca Raton : Taylor & Francis, 2017. | "A CRC title." | Includes bibliographical references and index.
Identifiers: LCCN 2016016878 | ISBN 9781498742658 (alk. paper)
Subjects: LCSH: Silicon solar cells. | Photovoltaic power generation. | Polycrystalline semiconductors. | Solar cells.
Classification: LCC TK2960 .S67 2017 | DDC 621.31/244--dc23
LC record available at https://lccn.loc.gov/2016016878

Visit the Taylor & Francis Web site at
http://www.taylorandfrancis.com

and the CRC Press Web site at
http://www.crcpress.com

Contents

Preface

Terrestrial photovoltaics gained full credibility more than 50 years ago, with the paper by Shockley and Queisser [1], setting the physical limits of conversion efficiency for single-junction solar cells. More than half a century of systematic R&D was necessary for photovoltaics to build up a sound economic platform and reach an industrial scale foreseeing 400–500 GW cumulative global installed capacity for 2020, and at least 50% of worldwide electric power generation for 2050.

As solar cells are predominantly made of crystalline silicon, the practical availability of a low-cost silicon feedstock emerged half a century ago as a prerequisite to success. The initial assumption in favor of silicon as a semiconductor material to solar cells, enunciated in the absence of any experimental evidence, predicted that solar cells could operate with sufficient efficiency also using silicon substrates of lesser quality than electronic grade silicon. The challenge was, therefore, to define an appropriate solar silicon quality and to develop industrial processes capable of replacing the Siemens process, which was already supplying the purest grade of silicon for solid-state semiconductors, but at a cost noncompatible in the long term with massive deployment of photovoltaics in terrestrial applications.

Three main options were originally retained for replacing the Siemens process.

The first was the direct upgrading of commercial, metallurgical grade silicon (MG-Si) by pyro-metallurgical and physical processes, addressing in the first instance the removal of lifetime killer impurities (mainly metals of transition elements).

The second was a further improvement to the first option adding to the preparation of MG-Si, the selection of pure raw materials (quartz and carbonaceous reductants) and a suitably clean operation of the metallurgical plant, thus also addressing the removal of boron and phosphorus, the respective main acceptor and donor elements to silicon.

The third was the gas phase purification of silicon using gases such as chlorosilanes or silane but replacing the Siemens bell jar-type reactor with less energy consuming and more productive reactors (free space or/and fluidized-bed reactors [FBRs]).

In all cases, the final feedstock had to be compatible in quality with the fabrication of high efficiency solar cells, meaning detailed knowledge about the effect of dopants, recombining impurities, and structural defects in the basic substrate.

The three options required advanced contributions in materials science and semiconductor physics, along with the optimization of process schemes and the development of analytical procedures able to detect and quantify impurities at the level below part per billion of atoms or less.

The adventure is hopefully not yet at its ultimate stage, but already presents the potential to bring photovoltaics to a leading position as an energy resource for our world.

The progress of photovoltaics, through more than half of a century, passed through a number of temporary crises of a political, financial, and industrial nature, as shown in Chapters 4 and 5. These chapters take a closer look at the history of companies and give illustrative examples of how PV development was indeed dramatically influenced by socio-political changes in Europe and in the world.

The aim of this book, however, is first of all, to give a rational appraisal of the numerous R&D activities carried out in the framework of low-cost silicon feedstock programs, generally sponsored by national and international agencies. Among them, the Japanese NEDO, the US DoE, and the European Commission played a major role, as direct evidence is shown in all the following chapters.

Advanced gas phase polysilicon production processes, for example, the FBR concept, are still making impressive progress in decreasing costs as Chapter 5 shows. Some projects at the industrial scale and others still at pilot scale present the promise to reduce the cost of silicon down to $12/kg [2]. This would set a new limit for the cost of high purity silicon.

Advanced gas phase processes are now in competition with processes working on different conceptual frames, aiming at the production of solar grade silicon directly from MG-Si as described in Chapters 2 and 3. Among them, the Elkem process is the most accomplished industrial example, with an annual capacity of 6500 t/year as described in detail in Chapter 4. Following another metallurgical concept, the Canadian company Silicor has announced the construction of a plant for the production of solar silicon in Iceland, to be ready in 2018, with a production capacity of 16,000 t/year at a projected cost of $9/kg [3].

Finally, Chapter 6 covers the fundamental work carried out by Kasuki Morita and colleagues in the framework of Japanese research activities addressed at solar silicon development.

The authors of this book have all actively worked for a long time in the solar silicon field and have a profound common belief in the final success of PV. Prospectively, this success is necessary because of the finite availability of natural resources, but it is also possible thanks to the infinite supply of sun power, at least in the perspective of humanity. The authors are, therefore, truly qualified to share their critical view on the problems encountered in R&D activities on the way to solar silicon, and to suggest some solutions for the future.

We hope to have shown, also, that the R&D activities carried out in this field were and still are an outstanding contribution to the advancement of materials science.

As leading editor, I am particularly indebted to Bruno Ceccaroli and Eivind Øvrelid for having shared the editorial work with me, as well as to Simona Binetti, Bill Brenemann, Tonio Buonassisi, Carlos del Cañizo Nadal, Yves Delannoy, Matthias Heuer, Stein Julsrud, Kasuki Morita, and Ragnar Tronstad, who contributed and worked hard for the success of this book.

My personal gratitude goes also to Dr. Wolfgang Palz, the promoter of an invaluable amount of R&D projects as director at the European Commission.

Sergio Pizzini

REFERENCES

1. W. Shockley and H. J. Queisser, 1961. Detailed balance limit of efficiency of p–n junction solar cells, *J. Appl. Phys.*, 32, 510–519.
2. http://www.pv-magazine.com/news/details/beitrag/pilot-production-begins-on-daqos-150-millionramped-up-polysilicon-capacity_100020000/#ixzz3eYpjSL4I
3. http://www.pv-tech.org/news/silicor_materials_selects_contractor_for_low_cost_iceland_poly_fab

Editors

Bruno Ceccaroli earned a docteur ès sciences degree in nuclear chemistry from the University of Strasbourg (1981, France). He graduated candidatus realium from the University of Oslo (1978, Norway). He started his industrial career at Saint-Gobain Recherche in Paris-Aubervilliers (1981–1985, France) dealing with physical deposition of thin films on glass and polymer substrates. He joined the Norwegian metallurgical group Elkem in 1986, being first located at the Bremanger plant (western Norway, 1986–1991) and then the Fiskaa plant in Kristiansand (southern Norway, 1991–1999). He served successively as development manager, production manager at the Silgrain department (Bremanger), plant manager (Fiskaa), industrial manager, and director for downstream strategy in the silicon metal division until he left the company in December 1999. During his involvement with Elkem (1986–1999), he played a central role in the solar grade silicon program using the company's proprietary metallurgical process. In 1998–1999, he served as director of the board to the newly incorporated ScanWafer AS, which produced multicrystalline silicon wafers for the solar market. In 2000, he joined the Renewable Energy Corporation (REC group) serving as CEO for ScanCell AS (the subsidiary of REC dedicated to manufacturing solar cells using wafers from its associate company ScanWafer). Within the REC group, his focus switched back to silicon feedstock when REC joined forces with ASiMI (USA) and Komatsu (Japan) to develop the fluidized bed reactor process to granular polysilicon from silane gas. He served as director of the board (2002–2005) to the joint venture Solar Grade Silicon (SGL LLC) and director of research at REC Silicon LLC (2005–2007). Since 2008, he has been entirely dedicated to his own company, Marche AS, a consultancy and investment company. He provides advice on strategy to boards and executives for various global companies participating and/or investing in materials and photovoltaic industries. He is also the founder of several advanced technology new ventures (e.g., n-Tec AS for the production of carbon nanotubes and Isosilicon AS for the enrichment of silicon and other light isotopes), which continue to keep his attention and funding. He is the author or coauthor of several review articles and book chapters on silicon purification for solar cells.

Eivind Øvrelid (born February 2, 1967) is a research manager for the PV silicon department in Sintef Materials and Chemistry and an associate professor at NTNU (The Norwegian University of Science and Technology) within PV materials. He earned a PhD in materials technology. In 1997, he started to work as a scientist for Sintef Materials in the area of light metal refining. In 1998, he moved to Glomfjord in the northern part of Norway and worked as product development manager and later R&D manager at Sinor, a company producing single crystals for the Semi and later the PV market. In 2003, he returned to SINTEF and was responsible for the laboratory for directional solidification. During the period 2003–2008, he was involved in the development of the Solsilc process. He became the research manager in 2012. Eivind Øvrelid is the author or coauthor of 70 papers.

Sergio Pizzini is a full professor in physical chemistry, and earned a doctorate in chemistry and a PhD in electrochemistry in 1966 at the University of Roma, Italy.

His professional career started at the Joint Research Centre of the European Commission in Ispra (Italy) and later in Petten (the Netherlands). Within this framework, he carried out innovative studies dealing with the electrochemistry of molten fluorides, including the development of a standard hydrogen electrode in molten fluorides. After leaving the Commission at the beginning of 1970, he joined the University of Milano as an associate professor, where he started a systematic investigation of solid electrolytes. Still maintaining his position at the university, he held the position of director of the Materials Department at the Corporate Research Centre of Montedison in Novara during 1975–1979, with activity in advanced materials for electronics and, later, as the CEO of Heliosil, where he developed a process for the production of solar grade silicon and patented a furnace for the directional solidification of silicon in multicrystalline ingots. In the year 1982, he had been solely with the University of Milano, and later, the University of Milano-Bicocca, as a professor of physical chemistry. Since November 2008 he is a retired professor, but is still associated with this University.

During this last period of activity, in addition to his teaching and management duties, he carried out systematic studies on semiconductor silicon, mostly addressing the understanding of electronic and optical properties of point and extended defects of Czochralski, multicrystalline, and nanocrystalline silicon in the frame of national and European Projects, using EBIC, LBIC, photoluminescence, EXAFS, and Raman spectroscopies.

Sergio Pizzini is the author of more than 250 scientific papers published in international peer-reviewed journals. He authored or coauthored three books, two titled *Advanced Silicon Materials for Photovoltaic Applications* and *Physical Chemistry of Semiconductor Materials and Processes* have been recently published by Wiley & Sons Ltd., and the third, coedited by G. Kissinger, titled *Silicon, Germanium, and Their Alloys: Growth, Defects, Impurities, and Nanocrystals*, published by CRC Press.

Contributors

Simona Binetti
Department of Materials Science
University of Milano-Bicocca
Milano, Italy

William C. Breneman
Technology, Research and
    Development, REC Silicon, Inc.
Moses Lake, Washington

Tonio Buonassisi
MIT Photovoltaic Research Laboratory
Boston, Massachusetts

Bruno Ceccaroli
MARCHE AS
Kristiansand, Norway

Yves Delannoy
Grenoble-INP SIMaP Lab
EPM Grp
St Martin d'Heres, France

Matthias Heuer
Calisolar GmbH
Berlin, Germany

Stein Julsrud
Technology, Research and Development,
    REC Silicon, Inc.
Moses Lake, Washington

Kasuki Morita
Department of Materials Engineering,
    Graduate School of Engineering
The University of Tokyo
Tokyo, Japan

Carlos del Cañizo Nadal
Solar Energy Institute
Polytechnic University of Madrid
Madrid, Spain

Eivind Øvrelid
SINTEF Materials and Chemistry
Department of PV Silicon
Trondheim, Norway

Sergio Pizzini
Department of Materials Science
University of Milano-Bicocca
Milano, Italy

Ragnar Tronstad
ELKEM AS, Technology
Kristiansand, Norway

1 Purity Requirements for Silicon in Photovoltaic Applications

Carlos del Cañizo Nadal, Simona Binetti, and Tonio Buonassisi

CONTENTS

1.1 INTRODUCTION

Microelectronics measures its progress in terms of the number of transistors per chip, decrease in channel length, or supply voltage, but normally takes for granted the fact that it relies on the capability of silicon producers to purify silicon to extreme grades. Electronic-grade (EG) silicon, in fact, is a material of 99.9999999% (9N, i.e., nine nines), or even 99.999999999% (11N) purity, in a process that consumes hundreds of kWh per kg, and still needs further processing to grow crystalline ingots and slice them into wafers. This purity figure is exciting as it concerns a material with just one impurity atom every billion or hundred billion atoms, produced in an industrial environment. Microelectronics can afford the cost of this process because the share of the purification process cost in the final cost of a chip is insignificant, as it amounts to less than 0.5%, on the base of a rough estimate made by considering an annual market of 30,000 tons of silicon at US$50/kg, in a global microelectronics business of US$300 billion [1].

The photovoltaic (PV) industry, on the other hand, is strongly interested in a cheap silicon feedstock, as silicon represents around the 10%–15% of the total module cost [2,3,4]. In the race toward competitiveness in the energy market, PV aims at the cheapest pure silicon that does not compromise module performance.

This concern was present in the PV field from the early times of the technology, and special R&D efforts were made after the oil crisis of the 1970s, in particular in the United States through a large Department of Energy research program. Research was conducted under the general assumption that a purification process for the solar industry should not be as demanding as that for the semiconductor industry, in order that any reduction in the production cost could be worth it, even at the expense of a "dirtier" and "more-defective" material. A number of very different routes were followed, and a large amount of knowledge was accumulated in that period (see, e.g., [5,6] and Chapters 2 and 3 of this book).

But the progress of PV technology did not meet the expectations of the moment, PV was relegated to the funding priorities of the R&D supporting agencies, and no practical innovation in the Si-purification field was finally transferred to the PV industry, although some of the results were further applied for the semiconductor business [7].

The PV companies survived in the absence of a dedicated Si-purification process because their market was rather small, and they could benefit from the off-spec material and the idle capacity of the industry at reduced prices. The PV industry demanded around 3000 metric tons of purified silicon in 2000, to be compared to the 17,000 tons devoted to microelectronics [8]. However, the rapid growth of PV in the first 2000 decade quickly changed terms, increasing the demand of ultrapure silicon for PV to levels where the off-spec quantity or the extra-capacity was not enough. In 2006, the silicon demand for solar surpassed that for electronics for the first time, and by the end of the decade around 80% of the 100,000 tons produced by the industry were dedicated to the PV market [9]. Silicon suppliers were not able to attend the growing PV demand, so that a silicon shortage was experienced in the second half of the 2000s, with prices rocketing to the hundreds $/kg level. Moreover, when the silicon players, both incumbents and newcomers, finally reacted to provide the material demanded by the PV industry, the market did not grow as much as it was expected and the scenario quickly changed to that of oversupply. Prices fell dramatically to the $20/kg range. This price level presses silicon producers to further reduce their production costs if looking for a sustainable business [10,11], bringing to the forefront again the question of what is the optimum balance between purity requirements and process complexity.

1.2   BACK AND FORTH THE SILICON VALUE CHAIN

Producing ultrapure silicon is not the end of the hystory, as the material needs further processing to become a solar cell. This processing involves a number of high-temperature steps, not only with the potential incorporation of contaminants, but also with the opportunity to rearrange the impurities already there. Another aspect that should be considered as "part of the purity" is the presence of structural defects, which break the perfect covalent-bonded arrangement of the semiconductor atoms and interact with the impurities and charge carriers.

The solar-grade (SoG) quality, therefore, is the final result of a series of process steps, and to define a SoG quality one has to take into account the role of each process step (i.e., of the value chain of crystalline silicon PV technology) on the final quality of the material, not simply the residual impurity content, as done conventionally [5,12].

Figure 1.1 reports schematically the value chain of crystalline silicon PV technology, from quartz to the solar cell, indicating the main effects in terms of contamination, purification, and defect creation. Note that the interconnection and encapsulation of cells in a PV module are not included, as they typically involve much lower temperatures that do not appreciably change the impurity content and distribution in the solar cell.*

A brief description of each step is reported in what follows.

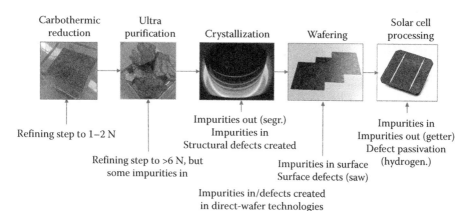

FIGURE 1.1   The PV crystalline silicon value chain, indicating the main effects in terms of impurity segregation and defect incorporation of each step.

---

* Actually, they do change them, as there are impurities that diffuse fast or moderately fast even at ambient temperature, and can be the cause of some detrimental effects at the module level [13,14].

TABLE 1.1
Typical Impurity Concentration Ranges in
Metallurgical Silicon

Impurity	Concentration Range in ppm(w)
C	50–1500
O	100–5000
Total metals	300–30,000
P	5–100
B	5–70

Source: Adapted from P. Hacke et al., 2010. *IEEE Photovoltaic Specialist Conference*, pp. 244–250.

1.2.1 METALLURGICAL SILICON PRODUCTION

Metallurgical silicon (MG-Si) is the raw material for pure silicon production. It comes from the carbothermic reduction of quartz at high temperature (~2000°C) in an electrical arc furnace (see Chapter 2 for details). Liquid silicon is produced, refined in a ladle, poured in a mold where it solidifies, and crushed. It contains from 1% to 3% of impurities depending on the raw materials and the type of electrode [15].

The main impurities typically present are C, O, Fe, Al, Ca, Ti, Mg, B, and P, whose concentration range is reported in Table 1.1 [16]. A careful selection of raw materials can lower the impurity content.

1.2.2 ULTRA PURIFICATION OF SILICON

To purify MG-Si, a three-step process known as the "Siemens process" is conventionally performed. MG-Si reacts with hydrogen chloride in a fluidized-bed reactor to synthesize a volatile silicon chlorides mixture, from which the trichlorosilane (TCS) (see details in Chapter 5) can be fractionally distilled in a number of columns, and then deposited as solid silicon by chemical vapor deposition (CVD) on slim Si seed rods heated by the Joule effect. Purities in the range of 9–11N (99.9999999%–99.999999999%) are achieved but at the cost of high energy consumption (in the range of 100 kWh/kg) and low efficiency deposition (for each mole of TCS converted to solid Si in the CVD reactor, 3–4 moles are converted to silicon tetrachloride) [16]. The impurities are removed from the MG-Si in the course of conversion to chlorosilanes, and it is important to underline that with the Siemens process the content of dopants is easily reduced at least three orders of magnitude. Special care is taken to avoid contamination during the deposition, the harvesting, and the post processing of the purified material. Crushing and packaging, for example, is typically done in a clean room environment with special equipment.

This ultrapure Si is usually called "polysilicon," as small Si crystals of different orientations form it, or "Si feedstock," as it is the input material that will be transformed into ingots and wafers. It is used for both the solar and the semiconductor industries. The latter demands the highest purity levels, specifically an extremely low content of B and P that should be lower than 20 ppba, while the former can relax the specifications and targets purity levels between 6N and 9N, depending on the solar cell type.

Si feedstock for solar application can also be produced by alternative processes to the Siemens one, as will be later described in Chapter 5 of this book. This is the case, for example, of the polysilicon produced by the deposition of monosilane in a fluidized-bed reactor, or the so-called upgraded metallurgical silicon (UMG-Si), in which several metallurgical-type purification steps are performed, each taking care of certain groups of impurities (metals, dopants, and light elements) [15] (see Chapter 3). In these Si feedstock technologies, the impurity content can vary over a wide range, as reported in Table 1.2 [11].

This lower-cost lower-quality UMG-Si material has been sometimes called "solar silicon," to distinguish it from the conventional "polysilicon" and to highlight that it is produced for PV applications. Nevertheless, we find this attribute somewhat restrictive, as the majority of the silicon devoted to solar nowadays comes from the conventional polysilicon route.

That is the reason why, to be more precise, the chapter authors propose to call "solar silicon" the purified silicon used as feedstock in the manufacturing of solar cells, whatever the process used to produce it, and whatever the level of purity achieved.

1.2.3 CRYSTALLIZATION AND WAFERING

This ultrapurified or purified material feeds the next step, which is the crystallization, where silicon is melted and solidified through a slowly controlled cooling, giving a monocrystalline or a multicrystalline (mc-Si) ingot.

In these solidification processes, dopant impurities are deliberately introduced to dope the material, typically boron for p-type Si and phosphorus for n-type, while unwanted impurities, often deleterious, can instead be also present in the final product, depending on the process conditions.

TABLE 1.2
Typical Purity Ranges for Different Ultra-Purification Routes

Impurity	Upgraded Metallurgical	Fluidized-Bed Reactor Deposition	Siemens Deposition (for Solar Industry)
P	300–1000 ppba	0.3–20 ppba	0.3–5 ppba
B	500–2000 ppba	0.3–20 ppba	0.1–5 ppba
Total metal	100–1000 ppbw	20–1000 ppbw	20–50 ppbw
C	50–200 ppma	0.5–10 ppma	0.25–1 ppma
O	100 ppmw	10–100 ppmw	0.5–5 ppmw

Source: Taken from G. Bye and B. Ceccaroli, 2014. *Sol. Energy Mater. Sol. Cells*, 130, 634–646.

In addition to impurity onset, the formation of crystal defects in a solidification process cannot be avoided in practice, as point defects (vacancies and self-interstitials) are equilibrium defects. These point defects can aggregate, due to stress and temperature gradients, producing voids and dislocations, as will be seen in Section 1.2.3.1. Also, bidimensional defects can form, like grain boundaries (GBs), when silicon solidifies in grains of different crystallographic orientations [17].

1.2.3.1 Monocrystalline Silicon for Solar Cells

The highest quality single crystals are produced by the float zone (FZ) technique, and record-efficiency solar cells have often been made with this FZ-Si material. Despite some industrial efforts developed in the past [18], up to now FZ silicon is too expensive to be used in PV industrial production, and most of the commercial monocrystalline silicon wafers for the PV industry are manufactured by the Czochralski (CZ) process.

From the impurity content point of view, the CZ process introduces mainly oxygen into the crystal lattice, which comes from the partial dissolution of the high-purity quartz crucible. Actually, CZ silicon is always a supersaturated solid solution of oxygen: the oxygen concentration at 300 K is around 18–20 ppma, equal to the solubility of oxygen in liquid silicon at 1412°C, that is, at the melting temperature of silicon.

In turn, this supersaturated solid solution of oxygen in Si is the origin of oxygen-related defects whose very nature depends on the thermal history of the material and on each thermal step occurring in the solar cell process.

A variety of oxygen-related defects, from clusters of few atoms called thermal donors to larger SiO_2 precipitates, can, in fact, be formed during the ingot cooling process or in the post-growth annealing step carried out on the wafers. The taxonomy of SiO_x complexes has been deeply studied for more than 30 years, and a reference work in this field has been written by Borghesi et al. [19]. The study of their effect on the minority and majority carrier properties has been extensively addressed as well.

CZ-Si ingots are also carbon contaminated: sources for carbon are predominantly the graphite parts in the furnace, such as the support crucible and the heaters. The C concentration is however below the solubility limit (<1 ppma), and has no important effect on CZ silicon-based solar cell performance.

Regarding crystal defects, nowadays CZ crystals are almost dislocation free, thanks to the use of a necking procedure. However, the control of intrinsic point defects (vacancies and self-interstitials) and their related microdefects (rings and voids) still remains a challenge and their presence can affect the efficiency of the devices, reducing it up to 1% absolute.

There are two main theories on point defect incorporation during the CZ silicon growth: Voronkov's theory [20,21] and a typical non-equilibrium theory, Abe's theory [22,23]. As the former is the most used in CZ-Si for PV applications, a short description of Voronkov's theory is reported here [20,21,24–28].

In Voronkov's equilibrium model, the concentration and distribution of point defects depend on the pulling parameter V/G ratio, where V is the growth rate and G is the temperature gradient. Voronkov's model states that growing a crystal at $V/G >$ critical ratio results in the incorporation of vacancies while the self-interstitials are undersaturated (see Figure 1.2).

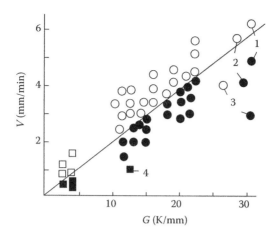

FIGURE 1.2 Observed microdefect type at various growth rate (V) and axial temperature gradient (G). ○: FZ-Si D-defects, □: CZ-Si D-defects, ●: FZ-Si A/B swirls, ■: CZ-Si A/B swirls. (Reprinted from the *J. Cryst. Growth*, 194(1), V. V. Voronkov and R. Falster, Vacancy-type microdefect formation in Czochralski silicon, 76–88, Copyright 1998, with permission from Elsevier.)

During the cooling, vacancies tend to agglomerate forming voids. When V/G is slightly higher than the critical ratio (thus for a low vacancy concentration), extended defects such as vacancy-oxygen complexes and oxygen precipitates are produced instead of vacancy agglomerates. For this reason, a narrow marginal band containing oxide particles, that is, where voids are generated, surrounds the main vacancy-containing region of a crystal.

If V/G is smaller than the critical value, silicon self-interstitials predominate, which later can agglomerate into interstitial-type defects (A/B swirls). As interstitial-type microdefects easily induce leakage current in solar cells, most silicon producers grow vacancy mode CZ crystals.

In addition to an inhomogeneous distribution of point defect clusters, it is often found that in CZ silicon impurities assume an inhomogeneous distribution pattern, quite similar to that characteristic of the point defect clusters. This distribution pattern takes generally the shape of striations, which have a radial symmetry. The impurity striations are caused by temperature fluctuations during crystal growth caused by the rotation of the crystal in a non-cylindrical thermal environment, or due to convection currents in the melt. If the striations are a result of dopant inhomogeneities or are due to carbon or oxygen local enrichment, they do not result in the generation of crystallographic defects during crystal growth. However, striations due to point defect clusters act as regions where nucleation of additional defects occurs when the wafers are annealed in subsequent steps, and could behave as lifetime killers.

Another detrimental defect for the solar cell performance is the so-called P-band ring: it consists of a narrow ring of grown-in oxide precipitates that can evolve toward an oxidation-induced stacking fault (OISF) ring after a thermal treatment [26]. The OISF ring usually occurs when the V/G is close to the boundary region but shifted a little toward the vacancy region. The P-band region is characterized by very low lifetime values, affecting the solar cells performance.

The fact is that any variation in the crystal-growth process parameters (pulling speed, melt convection, temperature profile, etc.) can lead to the formation of a pattern of defects and microdefects that can affect the minority carrier lifetime. Therefore, CZ wafers with comparable resistivity and impurity content can result in solar cells with different efficiencies according to the different density of grown-in defects. However, it has been recently shown that most of these defects can be partially annihilated with post-growth thermal treatments at relatively low temperatures (<300°C) [29].

As far as the metallic impurities* are concerned, any solidification process allows reducing their concentration from that in the feedstock, as they preferentially segregate in the liquid, rather than being incorporated in the solid Si phase during crystallization. Segregation coefficients of many transition metals are less than 10^{-4}–10^{-5}, enabling reduction of the metal concentration in the solid by two or three orders of magnitude with respect to the initial concentration in the melt, when 80% of the ingot has been solidified.

1.2.3.2   Multicrystalline Silicon for Solar Cells

Large grain polycrystalline silicon is the most commonly used material for solar cells, produced mainly by Directional Solidification (DS), a kind of a block casting technology that offers a significant cost reduction in comparison to the CZ pulling process. The 6–9 N purity polysilicon from different sources are used as feedstock in casting technologies.

Quartz (glassy silica) crucibles are generally used as containers. In the DS technologies a silicon nitride-based coating is deposited on the inside surface of the silica crucible in which silicon is melted and crystallized. This coating prevents wetting of liquid silicon on the crucible walls and thus the breaking of the crucible and ingot during the growth process due to the different thermal expansion properties of silicon and the crucible material.

On the other hand, this crucible coating is also the main source of nitrogen and of silicon nitride (Si_3N_4) precipitates that can be formed and included in the ingot [30].

DS produces a polycrystalline silicon conventionally named multicrystalline Si (mc-Si) with an oriented structure of columnar grains, where the grain size varies from some millimeters to centimeters. The grains are often twinned and sometimes heavily dislocated (dislocation densities between 10^4 and 10^8 cm^{-2}) [31]. In SoG mc-Si dislocations are mainly of 60° or screw type, dissociated into 90° and 30° partials in the case of 60° dislocations or in two 30° partials in the screw ones. Perfect Lomer dislocations can also be found as grown-in defects or can be formed from plastic deformation during the cooling-down process.

Oxygen and carbon are also the main impurities in mc-Si. The concentration of oxygen ranges from 5 to 10 ppma (much lower than in Cz-Si†), while the carbon

---

* Metallic impurities generally behave as lifetime killers.
† Due to the presence of the silicon nitride coating that prevents, different from the Cz process, the partial dissolution of the silica crucible.

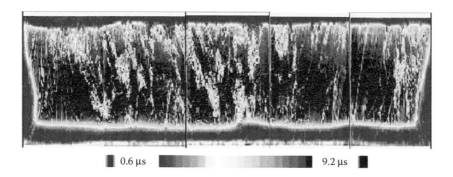

FIGURE 1.3 Longitudinal µ–PCD lifetime mapping of a mc-Si ingot. (Reproduced from the *J. Cryst. Growth*, 360(1), C. W. Lan et al., Grain control in directional solidification of photovoltaic silicon, 68–75, Copyright 2012, with permission from Elsevier.)

(coming from the graphite heaters and thermal shields) is always present in higher concentration up to the solubility limit (5–10 ppma). Carbon may nucleate in specific crystal sites and form carbide precipitates, potentially causing wire breakage during wafering processes, or electrical shunts in solar cells.

The DS process, as any other solidification process, benefits from the preferential segregation to the liquid phase of the metallic impurities. In comparison to the single-crystal-growth processes, the segregation coefficients of metal impurities are, however, higher in multicrystalline silicon growth process, due to the higher growth rates and to an enhanced segregation at GBs [32,33]. Eventually, the final metallic impurity profile along the ingot is also determined by the in-diffusion of impurities from the silicon nitride coating and from the portion of the ingot that solidifies last.

Although ultrapure silicon nitride (purity of more than 99.995%) is used for protective crucible coating, it is not only at the origin of the nitrogen contamination and of silicon nitride (Si_3N_4) precipitates, as already mentioned [30], but also the main source of metallic impurities, in particular of iron. Diffusion of impurities from the silicon nitride lining is responsible of a zone in the ingot characterized by reduced lifetime, the "red zone." Its width is comparable to the diffusion length of iron in solid silicon and ranges around 10–20 mm (see Figure 1.3) [30,34].

1.2.3.3 Direct Wafer Approaches for Solar Cells

The wafering process of a silicon ingot is carried out by wire saws and is responsible for large kerf losses, as will be seen in the next section. Direct wafer crystallization and ribbon technologies are examples of processes which have been studied and developed up to the industrial scale to reduce material losses due to the wafering step.

Among different ribbon technologies, even the most advanced, such as edge-defined film fed growth string ribbon, and ribbon growth on substrate) [36–38] did not succeed, however, at the industrial production stage, due to a number of challenges which have not yet been overcome.

One of the major drawbacks with ribbon silicon materials is their high disloca-tion density (up to 10^{10} cm^{-2}) and the generally higher impurity concentration as compared to silicon wafers obtained from directionally solidified mc-Si ingots. The specific puller design and the introduction of gases such as CO or CO_2 close to the meniscus during growth does significantly affect the impurity distribution in the bulk. At the same time, graphite dies introduce a large amount of carbon that is homogeneously distributed in the bulk and remains in high supersaturation condi-tions up to very high annealing temperatures [39,40].

Ribbon silicon may also contain small metallic precipitates due to fast crystal-lization temperature profiles [41].

1.2.3.4 Wafering

After cropping the ingots to prisms of square or pseudo-square cross section, the wafering step consists of slicing them in wafers 150–200 μm thick, using wire saws (with dramatic losses of material which is removed by an abrasive slurry).* Wafer surfaces are contaminated with residues from the slurry, and present a distribution of micro-cracks resulting from the stress of the abrasive process, which may propagate 5–10 μm into the wafer bulk [42]. Etching and chemo-mechanical processes can reduce the impact of these defective surfaces to a minimum.

1.2.4 SOLAR CELL PROCESSING

Solar cell processing includes the surface texturization, typically carried out by chemical means, the formation of the p-n junction, the antireflective coating (ARC) deposition, and the contact deposition and firing.

Each step is a potential source of contaminants, and high-temperature steps (mainly phosphorus diffusion, which is typically carried out at temperatures around 900°C) play a relevant role in the impurity evolution. Not only can they ease the in-diffusion and distribution of impurities in the bulk, but they can also help in their removal or neutralization, via "gettering processes" and "defect engineering approaches." The role of the anti-reflective layer deposition process, usually by plasma-enhanced chemical vapor deposition, is particularly important on the final quality of silicon, as it can induce "hydrogen passivation." These concepts will be further reviewed in Section 1.5.

1.2.5 PURITY REQUIREMENTS AS A FUNCTION OF THE FINAL CELL PERFORMANCE

As already anticipated in Section 1.1, due to the influence of the various process steps required to make a solar cell, the problems arising when attempting a definition of the purity requirements for a "solar silicon" are now apparent.

The same Si feedstock may give different cell efficiencies depending on the crys-tallization, wafering, and cell processing steps. Impurity segregation in substrates

---

* Today diamond wire slicing is also used, which avoids the use of abrasive slurries and further contami-nation of wafers.

CZ-grown will be different from that of DS-grown substrates, since the presence of crystal defects will change the dynamics of the impurities segregation during a thermal step in the cell process, and favor the trapping of impurities, and their further evolution as defect complexes and precipitates.

Because of this "cascade" effect, purity requirements for solar silicon should be defined having as a reference the target of solar cell performance, in particular the solar cell efficiency. This means that the impurities present in the final solar cell should be tracked back, following their evolution and origin throughout the previous steps in the value chain until the Si feedstock manufacturing, and then set the maximum acceptable contamination level.

Note that this maximum contamination level will be in general different for different impurities, as their properties differ largely (liquid–solid segregation, diffusivity, solubility, mechanism of nucleation, etc.), and so will their evolution during the whole Si process. Extensive experimental and modeling work concerning the impurity content and solar cell efficiency is present in the literature, as will be shown in Section 1.4. When accounting for these literature results, this "value-chain" logic should be present in order to have a clear idea of what step the reported impurity concentration refers to, and to what device performance target it is compared with.

1.3 QUALITY GOALS DEPEND ON CELL ARCHITECTURE

The influence of impurities and defects on the cell performance will also depend on the solar cell architecture, so that the same impurity content and distribution in a substrate at the end of the solar cell process can result in different solar cell efficiency degradation levels for different cell structures.

In the studies analyzing the maximum allowable efficiency for single-junction crystalline Si solar cells, normally a perfect crystal structure with no unintentionally introduced impurities is taken as a reference [43,44–46]. This assumption, together with other idealizations and accounting for thermodynamic constraints, leads to limiting efficiencies in the range of 30%.

From there, practical limitations should be taken into account that further reduce the maximum achievable efficiency for a silicon solar cell, as Figure 1.4 shows, bringing the limit to the 26% range [47]. It has to be noted that this analysis still considers a high-quality substrate, corresponding to contaminant impurities below a concentration of 10^{12} cm^{-3} (i.e., purity levels higher than 10 N).

The presence of impurities and defects will mainly manifest in an increase in the recombination of minority carriers through traps, following the Shockley–Read–Hall (SRH) model. This recombination mechanism is typically dominant in the bulk of a semiconductor, as compared to the radiative or Auger components of the recombination itself.

How detrimental the defects and impurities are depend on the weight of the SRH recombination term relative to others, specifically to those associated with surfaces and contacts. The recombination occurring at surfaces, contacts, and bulk will be different in different cell architectures, and hence they can be affected differently by the same impurity level.

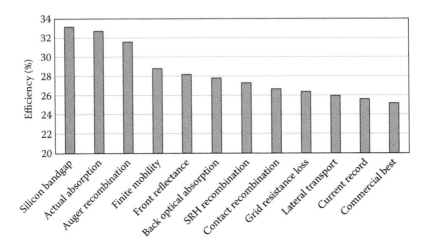

FIGURE 1.4 The waterfall loss in going from the thermodynamic limit to practical cells. (Adapted from R. Swanson, 2005. Approaching the 29% limit efficiency of silicon solar cells, in 31st *IEEE Photovoltaic Specialist Conference*, Lake Buena Vista, Florida, pp. 889–894.)

As an example, if passivation at the surfaces is maximized, solar cell performance will be very sensitive to the wafer quality,[*] but if the cell is limited by the recombination at the front emitter the quality requirements can be relaxed to a great extent. Note that this does not mean that the latter case is preferable: for the same purity level, in the latter case the maximum achievable efficiency will be lower than in the former because the term corresponding to recombination due to traps will sum up with a major contribution due to recombination at the front of the cell. Therefore, when emitter recombination is dominant for a specific cell architecture, having a high-purity substrate will not make a difference with respect to having a low-purity one.

SRH recombination is typically expressed in terms of lifetime τ_{SRH}, a statistical term indicating *the average time that it takes for an excited carrier to recombine*[†]: the higher the recombination, the lower the lifetime. As a first approximation, the SRH lifetime is inversely related to the concentration of defects through the equation

$$\frac{1}{\tau_{SRH}} = \sum \frac{1}{\tau_{SRH,defect}} = \frac{1}{v_{th}} \sum \frac{1}{\sigma_{defect} C_{defect}} \tag{1.1}$$

with v_{th} the minority carrier thermal velocity, σ_{defect} the minority carrier capture cross section, and C_{defect} the concentration of active impurities and defects.

---

[*] In terms of impurity content.
[†] The diffusion length L_D, or the average length that an excited carrier travels before it recombines, can equally be used as the figure of merit, as it is proportional to the square root of lifetime.

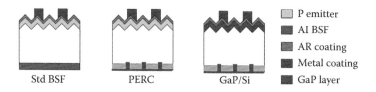

			☐ P emitter
			■ Al BSF
			■ AR coating
			■ Metal coating
Std BSF	PERC	GaP/Si	■ GaP layer

FIGURE 1.5   Cell architectures selected for the efficiency versus lifetime simulations.

The SRH lifetime is then a good indicator of material quality, so that a good way to estimate the impact of the purity level in a specific cell architecture is to develop a precise device model of the cell, and sweep the results in SRH lifetime.

The simulation will be done below for three different cell architectures, schematically presented in Figure 1.5. Here, "Std BSF" indicates a cell with a conventional P front emitter and a rear full-area Al BSF (back surface field); "PERC" (passivated emitter and rear cell) indicates a cell with an improved P front emitter and a rear passivated surface locally contacted through holes; and "GaP/Si" is a cell with the same rear structure as the PERC and a hetero-emitter at the front, made of n-doped GaP. The first two represent industrially proven architectures [48], while the last one is a proposal for an advanced cell architecture that builds on the experience of rear passivation in PERC structures and carrier-selective contacts in heterojunction designs [49].

The Si substrate is in the three cases p-type, 1.79 Ω cm and 180-μm thick. The GaP/Si structure has also been simulated for a 80-μm-thick Si substrate, to show how the wafer thickness can influence cell performance.

The three cell architectures have been modeled with Sentaurus Device,* selecting representative and widely accepted input parameters to describe the different cell architectures. Details can be found in References 50 and 51, and references therein.

Figure 1.6 shows the cell efficiency when the bulk lifetime goes from 1 to 5000 μs in the four cases. Two regions can be clearly identified in the curves: for low bulk lifetimes the cell efficiency strongly depends on the lifetime, as the material quality is limiting cell performance, while for high lifetimes the material quality is no longer limiting and efficiency saturates.

The threshold values differ in the four cells, depending on the balance of the various recombination terms. The better the surface passivation is, the more sensitive the structure is to material quality, making the threshold value higher, which is observed for the sequence Std BSF, PERC cell, and GaP/Si cell. In the case of the thin cell, the photoexcited carriers need to travel shorter distances to be collected, making the threshold value decrease as compared to that for the thick device. Note that the maximum efficiency is lower for the thin device due to a reduction in the light absorption and hence in the photogeneration yield.

Given a substrate with a certain concentration of impurities and defects at the end of the cell process, which translates in a certain lifetime, Figure 1.6 indicates the efficiency that can be achieved, depending on the cell architecture. But a relevant aspect

---

* *Sentaurus Device* is a multidimensional device simulator from Synopsys® capable of simulating electrical, thermal, and optical characteristics of semiconductor devices, in particular silicon-based ones.

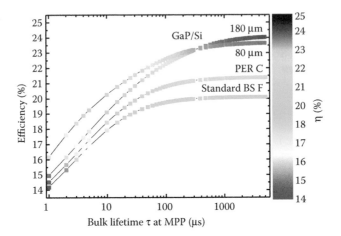

FIGURE 1.6 Efficiency as a function of the bulk lifetime at the maximum power point (MPP) during solar cell operation for different device architectures. Color coding is added corresponding to the y-axis for easier comparison with simulations in Figure 1.8. (J. Hofstetter et al., Materials requirements for the adoption of unconventional crystal-growth techniques for high-efficiency solar cells. *Prog. Photovolt. Res. Appl.* 2016, 24, 122–132. Copyright Wiley-VCH Verlag GmbH & Co. KGaA. Reproduced with permission.)

that should not be forgotten, which is not explicit in Figure 1.6, is the injection-level dependence of the lifetime, as the following expressions show:

$$\tau_{SRH} = \Delta n \cdot \frac{\tau_{0,e}(p+p_1) + \tau_{0,h}(n+n_1)}{np - n_i^2} \tag{1.2}$$

with

$$n_1 = n_i \exp\left(\frac{E_t - E_i}{kT}\right) \tag{1.3}$$

$$p_1 = n_i \exp\left(\frac{E_i - E_t}{kT}\right) \tag{1.4}$$

$$\tau_{0,e} = \frac{1}{N_t \sigma_{t,e} v_e} \tag{1.5}$$

and

$$\tau_{0,h} = \frac{1}{N_t \sigma_{t,h} v_h} \tag{1.6}$$

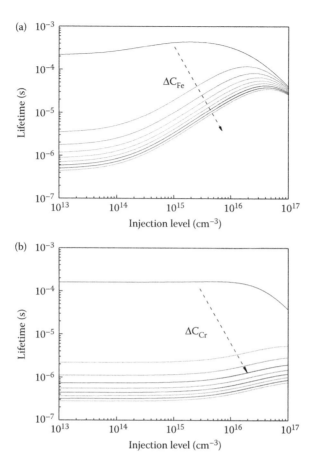

FIGURE 1.7 Total lifetime as a function of the injection level for Fe in p-type silicon (a) and Cr in p-type silicon (in the form of Cr-B pairs) (b). Higher lifetimes correspond to lower concentrations. ΔC_{Fe} and ΔC_{Cr} are the corresponding increase of Fe and Cr concentrations.

where Δn is the injection level (excess carrier concentration), n_i the intrinsic carrier concentration, E_t the defect energy (referred to the intrinsic Fermi energy), N_t the defect concentration, $\sigma_{t,e}$ and $\sigma_{t,h}$ the capture cross sections, and v_e and v_h the thermal velocities of electrons and holes, respectively.

The injection-dependent effect is illustrated in Figure 1.7 for two representative impurities (Fe and Cr) in p-type silicon. As the capture cross sections for electrons and holes are very different for Fe, lifetime changes steeply with injection level, which is not the case for Cr, for which the lifetime is almost independent of injection level until the intrinsic recombination mechanisms (radiative, Auger) come into play.

This injection-level dependence of the lifetime has relevant implications in the cell performance, see for instance Macdonald and Cuevas [52]. Here the attention is

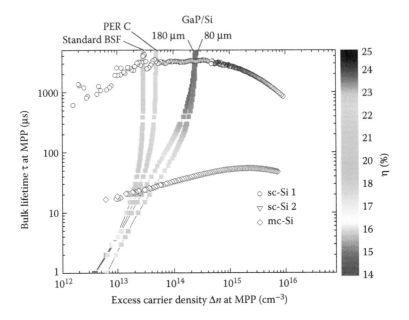

FIGURE 1.8 Bulk lifetimes at the maximum power point (MPP) as a function of the excess carrier density for the four cell architectures, with the corresponding efficiency indicated by the color coding. Open symbols are measured lifetime as a function of excess carrier density in two different wafers, a single crystalline (circles) and a multicrystalline one (diamonds). (J. Hofstetter et al., Materials requirements for the adoption of unconventional crystal-growth techniques for high-efficiency solar cells. *Prog. Photovolt. Res. Appl.* 2016, 24, 122–132. Copyright Wiley-VCH Verlag GmbH & Co. KGaA. Reproduced with permission.)

in the fact that it is the lifetime value at the injection level where the cell is operating which characterizes the material quality influence on the cell performance.*

To make this explicit the bulk lifetime at the maximum power point (MPP) is reported in Figure 1.8 as a function of the excess carrier density (i.e., the injection level) for the four cell types, and the efficiency is given for each pair lifetime–injection level [51,53], with a color coding which is the same as in Figure 1.6

Plotted in Figure 1.8 are also two injection-dependent bulk lifetime curves, corresponding to a single-crystalline wafer and a multicrystalline one. From the crossover point between the simulated curves at the MPP and the injection-dependent lifetime curves, the PV conversion efficiency potential of the respective material can be read. For the monocrystalline example, efficiencies of 23.6%, >22%, and 20% are within reach for the GaP/Si, the PERC, and the Standard BSF cases, respectively; for the multicrystalline case, results are 21.4%, 19%, and 18.8%, respectively. This kind of assessment can be made for any wafer, as long as its injection-dependent lifetime is measured. Note that for a proper analysis, the measured lifetime is that at the end of the solar cell process.

---

* In particular, the injection level at the maximum power point (MPP), if we refer to the cell efficiency.

A more rigorous analysis should take into account aspects such as the spatial variation of the injection level, and the type and concentration of the lifetime-limiting defects used for the solar cell simulation. These aspects are discussed in detail in Reference 51, where a more broadly applicable version of Figure 1.8 is also presented. It also includes an example of how this methodology has been applied to guide the development of a new epitaxial silicon technology, which has succeeded in reaching lifetimes high enough to give devices over 20% efficiency.

1.4 WHAT IS IMPACTING LIFETIME

SRH lifetime, as already mentioned, is determined by the impurities in the wafer, the crystal defects, and the interactions between them. It can be useful to review with some detail how knowledge of the impact of impurities and defects in solar cells has evolved to date.

1.4.1 ROLE OF DISSOLVED METAL CONTAMINANTS

Pioneering research was conducted in the 1970s (for instance, Wakefield et al. [54]), but the reference work in the field is the "Westinghouse study," sponsored by the Jet Propulsion Laboratory. It included the growth of nearly 200 monocrystalline ingots with controlled introduction of selected impurities, the processing of more than 10,000 solar cells, and the establishment of characterization procedures and device-performance models for data analysis. The results were summarized in Reference 55 that displayed the "iconic" figures reproduced in Figure 1.9, which show the degradation in cell performance for p-base and n-base solar cells as compared to a non-contaminated reference solar cell efficiency of 14.1%.

The analysis of the experimental results leads to the conclusion that the dominant mechanism responsible of the cell degradation with the addition of impurities was the reduction of the diffusion length in the bulk (equivalently, the carrier lifetime), which was quantified assuming a homogeneous distribution of impurities. For Ni, Cu, and Fe the authors also observed substantial performance loss due to junction defect phenomena. Another interesting result was the realization that n-base devices are generally less affected by typical metallic impurities than are corresponding p-base devices.

More recently, an experimental effort was carried out in the framework of the European R&D project Crystal Clear on the effect of impurities in multicrystalline silicon ingots [56]. The study was restricted to a limited number of representative impurities (iron, chromium, nickel, titanium, and copper) that were intentionally added, in separate experiments, during the crystallization step. The non-contaminated reference cell efficiencies reached 15.5%. The degradation of the cell performance, in the case of iron, chromium, and titanium, is due to a reduction in the diffusion length, in the case of nickel mainly to an increase in the emitter recombination, and in the case of copper both to the increase of bulk and emitter recombination.

The experimental screening of impurity nature and concentration has undoubtedly an enormous value, but it requires a huge effort and is very expensive in terms

(a)

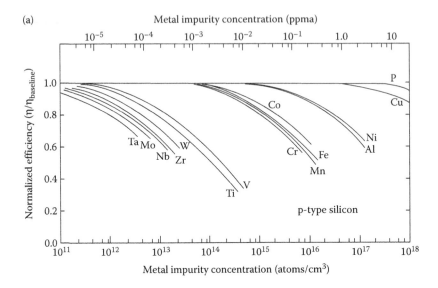

(b)

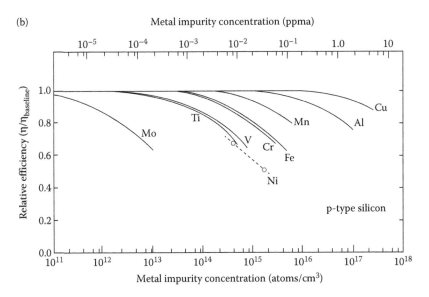

FIGURE 1.9 Solar cell efficiency versus impurity concentration for 4 Ω cm p-base devices (a) and 1.5 Ω cm n-base devices (b) as reported in Reference 55. (B. Ceccaroli and O. Lohne, Solar grade silicon feedstock, in *Handbook of Photovoltaic Science and Engineering*, 2nd ed. A. Luque and S. Hegedus, Eds., John Wiley & Sons, 2011, pp. 169–216, 2011. Copyright Wiley-VCH Verlag GmbH & Co. KGaA. Reproduced with permission.)

of time. Performing simulations is much faster and cheaper, and can give useful insight provided the simulations are based on well-established physical and chemical principles, and on reference values of the key impurity properties [57–59].

A recent example is shown in Figure 1.10, where the impact of the most common metallic impurities* in a number of cell architectures has been calculated assuming a matrix of dissolved impurities in single crystal material, in the absence of crystal defects [51]. Similar curves were estimated for Cr, Co, Fe, and Ni in a PERC structure in a previous work by Schmidt et al. [62], and the results match well.

The work of Schmidt et al. additionally compares their effect on p- and n-type cells, concluding that n-type material is less sensitive to Fe, not so much for Co, Cr, and Ni.

The work of Hofstetter et al. [51] expands the analysis to a wider set of impurities and allows us to compare the different sensitivities of the three cell architectures mentioned in the previous section to the impurity concentration measured at the end of the solar cell process. It can be seen that the more advanced architectures (PERC and GaP/Si) are more sensitive to the impurity content, that is, for the same concentration the degradation in cell performance is higher, as expected from cell structures that succeed in minimizing the recombination effects at the surfaces.

Note that in these simulations the impurity concentration refers to that measured at the end of the solar cell process, while in the experimental works of references [55,56] the impurity concentration is that of the as-grown material, before the solar cell processing.

1.4.2 METAL PRECIPITATION AT CRYSTAL DEFECTS

The role of impurity interactions with structural defects was first recognized by Pizzini et al. [63,64]. In Reference 63 multicrystalline ingots were contaminated with Fe, V, Ti, and Cr, and the experimental characterization was compared with a model based on the assumption of the absence of impurity–defect interactions (see Figure 1.11). The model predictions are fulfilled for Fe, V, and Ti concentrations below certain thresholds, but important deviations are observed for the case of Cr in the full range of the Cr concentrations explored, and above a threshold for all four impurities considered.

These results further challenged the hypothesis of a uniform distribution of dissolved impurities, by introducing inhomogeneously distributed structural defects as potential heterogeneous nucleation sites for metal precipitation. The combination of steeply falling solubilities and slowly falling diffusivities of metal impurities with temperature can easily lead to supersaturated solutions, inducing both defect reactions and precipitation during cooling and subsequent process steps [17,65]. This phenomenon is enhanced for some of these impurities, as they can form silicides that fit nicely into the Si lattice [66].

---

* The recombination properties of most of these metals have been recently determined by combining advanced characterization techniques (in particular, deep level transient spectroscopy measurements, DLTS [60]) and temperature- and injection-dependent lifetime spectroscopy (TIDLS) [61].

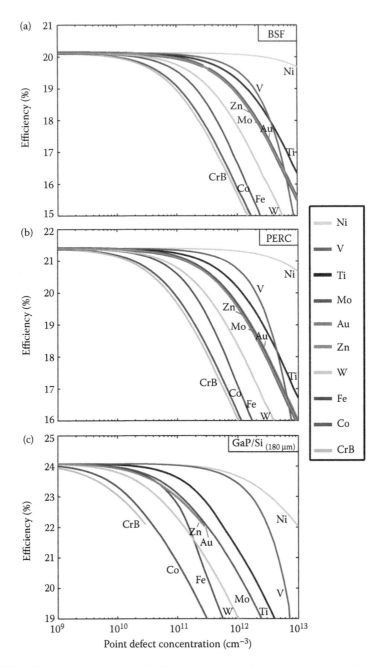

FIGURE 1.10 Simulated solar cell efficiency as a function of metal point defect concentration in a Std BSF device (a); a PERC device (b); and a GaP/Si device (c). (J. Hofstetter et al., Materials requirements for the adoption of unconventional crystal-growth techniques for high-efficiency solar cells. *Prog. Photovoltaics Res. Appl.* 2016, 24, 122–132. Copyright Wiley-VCH Verlag GmbH & Co. KGaA. Reproduced with permission.)

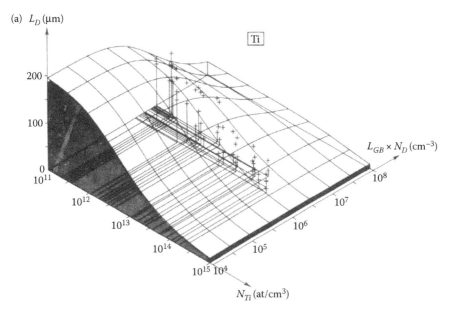

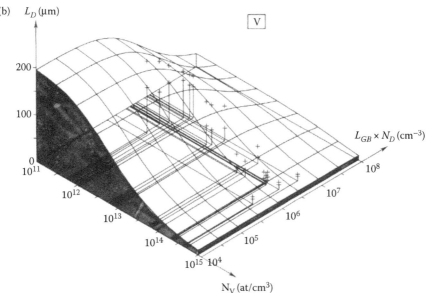

FIGURE 1.11 Experimental values of carrier diffusion length L_D [μm] as a function of the product $L_{GB} \times N_D$ (with L_{GB} [cm^{-1}] the average length density of grain boundaries and N_D [cm^{-2}] the average dislocation density) and the impurity concentration N_x in the as-grown material for Ti (a) and V (b). (Reproduced with permission of the Electrochemical Society, after S. Pizzini, L. Bigoni, M. Beghi, and C. Chemelli, On the effect of impurities on the photovoltaic behavior of solar grade silicon: II. Influence of titanium, vanadium, chromium, iron, and zirconium on photovoltaic behavior of polycrystalline solar cells, *J. Electrochem. Soc.*, 133(11), 2363–2373, 1986.) (*Continued*)

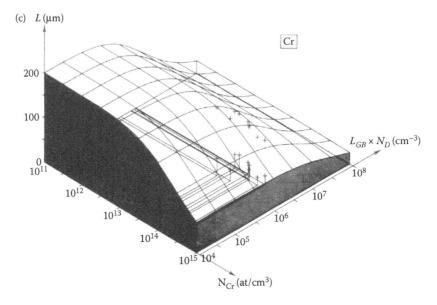

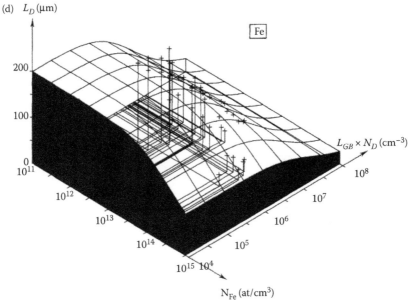

FIGURE 1.11 (Continued) Experimental values of carrier diffusion length L_D [μm] as a function of the product $L_{GB} \times N_D$ (with L_{GB} [cm^{-1}] the average length density of grain boundaries and N_D [cm^{-2}] the average dislocation density) and the impurity concentration N_x in the as-grown material for Cr (c) and Fe (d). (Reproduced with permission of the Electrochemical Society, after S. Pizzini, L. Bigoni, M. Beghi, and C. Chemelli, On the effect of impurities on the photovoltaic behavior of solar grade silicon: II. Influence of titanium, vanadium, chromium, iron, and zirconium on photovoltaic behavior of polycrystalline solar cells, *J. Electrochem. Soc.*, 133(11), 2363–2373, 1986.)

Recombination at a precipitate differs from that at a dissolved impurity, and several models have been developed to understand how [67,68]. The treatment of the metal precipitates in silicon as a Schottky barrier [69,70] was an inspiring hypothesis, which has been revisited recently [71]. Most importantly, the work by Pizzini et al. brought to discussion the hypothesis that much of the metals in multicrystalline silicon are present in the form of precipitates and not as point defects, and high-temperature annealing releases metals into the bulk. This hypothesis is now widely accepted, consequent to the results of neutron activation analysis [72] and synchrotron-based micro x-ray fluorescence (μ-XRF) [73] that confirmed the microprecipitates theory.

1.4.3 RECOMBINATION ACTIVITY OF DISLOCATIONS AND GRAIN BOUNDARIES

The aforementioned hypothesis concerning defects as nucleation sites for metal precipitation served to draw a more comprehensive and complex picture about the electrical activity of extended defects (dislocation and GBs) and their effects on solar cell performance, which include:

1. The intrinsic activity of these extended defects
2. The activity or the effect of precipitates and of the impurity atmosphere in the vicinity of the dislocation line or at the GBs

Despite more than 50 years of investigations, some aspects of the carrier recombination process at dislocations, namely, the origin of dislocation-related levels and those related to dislocation–impurity interactions, are still under debate. An excellent review can be found in Reference 74 and references therein.

The dislocations affect the lifetime of minority carriers, and the correlation between high density of dislocation and reduced lifetime in mc-Si is very well known. It has been described by fundamental recombination models [75,76], which have been more recently applied to explain performance limitations in PV-grade silicon materials [77,78].

Actually, it is recognized that clean dislocations should be electrically inactive or with a small recombination activity at room temperature [79]. It has been shown that glide dislocations introduced by plastic deformation and Frank-type partial dislocations formed around precipitates give rise to electron-beam-induced current (EBIC) contrasts in the temperature range lower than about 200 K. They become recombination inactive and their EBIC contrasts disappear at temperatures higher than 200 K [80].

Furthermore, it has been definitively proven that the recombination activity of a dislocation is enhanced by the segregation of impurities, a thermodynamically favorable process. Not only do transition metals introduce associated deep levels with a strong recombining activity [81], but oxygen and carbon segregation can also enhance the recombination activity of a dislocation [33,82,83].

As described in the previous section, this spontaneous interaction between dislocations and impurities results in a spatial inhomogeneity of the impurity distribution and this, in turn, results in the spatial inhomogeneities of the electrical and/or optical properties of the material, reflecting the dislocation distribution.

Besides the electrical properties, various other properties of a dislocation are greatly affected by the segregation or precipitation of impurities. For instance, the optical effects typical of a clean dislocation may be modified due to decoration with impurities. As a matter of fact, the dislocations are responsible for four luminescence bands (the D1–D4 lines) in the 0.8–1.0 eV range [84]. While the D3 and D4 are related to the dislocations themselves, the real origin of D1 and D2 is still unclear, but it has been shown that their features (energy positions and intensity) are affected by the presence of impurities like oxygen [85–89] and metals [90,91].

Recent developments of spatially resolved photoluminescence easily prove the local changing of lifetime in mc-Si associated to dislocations and dislocation–impurity phenomena, simply monitoring the dislocations-related luminescence bands [83,92,93] both in cut material and after the solar cell processes [94,95].

Also GBs in mc-Si, as a consequence of their highly defective nature, can behave as efficient recombination centers for photogenerated carriers, as well as effective sinks for impurities.

EBIC and electron backscatter diffraction studies have been extensively used to explore the impact of GBs and the impurity–GB interactions on the electrical properties of silicon devices. The relationship between GB geometrical character and its electrical activity have been studied for many years. Small-angle grain boundaries (SA-GBs) are known to be the most electrically active defects in multicrystalline silicon. This electrical activity has been associated to the edge-type and 60° screw dislocations located at the GBs and the higher the fraction of edge dislocations, the stronger the SA-GB electrical activity.

In a fundamental work of Chen et al. [96] it was shown that the recombination activity of clean GBs is in any case weak. However, it becomes stronger as the contamination level increases, and the recombination activity is clearly related to the GB character, as the random or high-Σ GBs* showed a stronger EBIC contrast than the low-Σ GBs. The relationship between the strength of the GB decoration and the GB structure was further confirmed by several authors [97].

These studies were enabled by the emergent ability to "see" precipitated metals and determine their chemical states, with the advent of synchrotron-based μ-XRF, μ-XAS, and high-resolution trasmission electron microscope (TEM) in the 2000s. It was confirmed that the higher the degree of disorder within the GB core structure, the higher the likelihood of metal interaction, the low solid solubility of metals in bulk silicon being a driver for impurity segregation/precipitation at structural defects [98]. This model was later extended to dislocations [99].

Further it has been found that GB activity can be reduced by hydrogen passivation, but, as expected, its effectiveness depends on the GB character and metal contamination [100,101].

It has also been shown [77] that the recombination activity of dislocations and GBs was strongly increased by a phosphorus emitter diffusion at about 900°C.

---

* The structure of grain boundaries is described using the concept of coincidence site lattice (CSL) and Σ is a CSL parameter.

On the other hand, the deposition of an hydrogenated silicon nitride antireflection coating layer significantly reduced the recombination activity of dislocations by hydrogen diffusion and passivation process. As atomic hydrogen in-diffusion has been reported to show less effective lifetime improvement as dislocation density increases [94], the beneficial effects of the antireflection coating deposition can vary, depending on dislocation density and decoration.

Recently, Bauer et al. [102] observed a direct correlation between the dark current losses in mc-Si solar cells and the density of Lomer dislocations at SA-GBs. This seems to demonstrate that perfect Lomer dislocations play a major role for recombination at subgrain structures in mc-Si, while the role of any impurities in the core of this type of dislocation in the recombination process is still under debate.

1.4.4 DOPANT-RELATED DEFECTS

1.4.4.1 B–O Pairs

Substitutional boron (B_s) is the most common p-type dopant in solar silicon, and interstitial oxygen (O_i) is the dominant impurity in CZ silicon, making the B–O pair the most known and studied dopant-related defect. This pair is responsible of reducing the minority carrier lifetime under illumination or carrier injection, with a consequent loss of up to 2% in the conversion efficiency of standard solar cells.

This effect is called light-induced degradation (LID), an effect that was observed for the first time in 1973 by Fischer and Pschunder [103], who recorded a degradation of the short-circuit current density and the open-circuit voltage of a CZ-Si cell during the first few hours of illumination, until a stable level was reached. They also proved that a 200°C annealing in the absence of illumination could completely recover the cell performance. From 1973 to 1997, the LID effect observed in CZ-Si solar cells was mainly explained with the formation of metal-containing defect complexes. In 1997, Schmidt et al. [104] proposed a defect reaction model involving only a B–O defect. This metastable defect was capable of explaining both the lifetime degradation under illumination as well as the recovery by a low-temperature annealing. The most important feature of this model is the formation of a defect pair composed of one interstitial boron and one interstitial oxygen atom (B_iO_i). Starting from 1998, a great deal of attention has been directed to understand the formation of this defect and how to suppress or avoid LID-related phenomena. The development of B–O complex theories up to 2003 is well summarized in Reference 105. In 2003, the experimental proof that B–O defect concentration increases monotonically with boron concentration and it is proportional to the square of the oxygen concentration allowed proposing that this defect could be a BO_{2i} complex [106]. In this model, fast-diffusing oxygen dimers O_{2i} are captured by substitutional boron B_s to form a B_sO_{2i} complex, acting as a highly effective recombination center. Bothe and Schmidt, in 2006, brought to a full development this theory [107]. In the following years, some experimental evidences ruled out this model, as it was shown that the LID process in compensated material is dominated by the net doping concentration instead of the absolute boron concentration [108], and also that the saturated concentration of defects depends on [B] and not on the starting concentration of O_2 dimers.

In 2010 an alternative model has been proposed by Voronkov and Falster, which is able to overtake some of the problems presented by the previous theories [109]. According to this model, the B-doped silicon already contains latent recombining defects, and the excess carriers induce an evolution of these latent centers into a recombination-active configuration. The latent defects are B_iO_2 complexes formed during the ingot cooling process, whose concentration is proportional to $[O]^2 \times [B_s]$. It is assumed that the B_i atoms are the result of the interaction of B_s with self-interstitials emitted by growing oxide precipitates or clusters. It is also assumed that the solubility of B_i species is small, inducing the precipitation of the excess B_i concentration under the form of nano-precipitates $[B_i]_n$.

The grown-in B_iO_2 defects are considered as "latent centers" (LC) of a low recombination activity, and behave as deep donor levels close to the mid-gap. The LID effect is supposed to be caused by the capture of excess electrons, arising from illumination, with subsequent reconstruction of LC into another atomic configuration of B_iO_2, the "slow stage recombination centre" (SRC), that is recombination active. This process is schematically depicted in Figure 1.12.

Nowadays the temperature/time for recovery by thermal annealing, used by many researchers, is 200°C for 30 min. The recovery can be made permanent by performing the annealing under simultaneous illumination [110]. LID can also be avoided by the use of p-type FZ, gallium-doped CZ, or n-type CZ substrates, as it has been experimentally confirmed that they do not show any performance degradation under illumination.

It must be said, however, that in the last few years several aspects of the light degradation process have been brought into question and yet a full understanding of the LID phenomenon has not been achieved.

Finally, LID was previously considered for CZ silicon wafers only because of their higher overall quality that made LID detrimental effects more evident. However, as

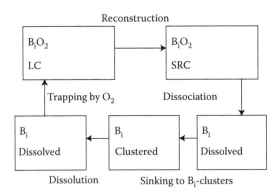

FIGURE 1.12 A flowchart showing the reconstruction of the LC defect upon illumination leading to the SRC active defect and cell degradation, subsequent recovery after thermal treatment under illumination, and final recreation of the latent defect in the dark. (Reprinted with permission from V. V. Voronkov and R. Falster, Latent complexes of interstitial boron and oxygen dimers as a reason for degradation of silicon-based solar cells, *J. Appl. Phys.*, 107(5), 53509. Copyright 2010. American Institute of Physics.)

the quality of multicrystalline wafers has been improved in the last decade due to optimized ingot and solar cell processes, LID effects are also turning more and more significant for mc-Si cell technology [111].

1.4.4.2 H–P or H–B Complexes

As reported earlier, the introduction of hydrogen in silicon during solar cell processing can passivate both dislocations and point defects. On the other hand, hydrogen can also induce the formation of H–B or H–P complexes. The structure of the acceptor-H and donor-H pairs in Si has been studied for many years using a variety of experimental methods and theoretical models (see Reference 112).

To our knowledge, the standard concentration of hydrogen is so low in silicon solar cells that these complexes have no effect on solar cell efficiency. However, if the use of hydrogen passivation is applied to compensated UMG silicon (see next section), or the H content of solar silicon is much higher than the standard one, the formation of these pairs should be taken into account, as they can affect the resistivity values.

1.4.4.3 Compensated UMG-Si

UMG and silicon feedstocks produced following other alternative purification routes (see Chapter 3) contain a high concentration of dopant impurities, which are difficult to remove with physical or chemical processes. A significant presence of both acceptor and donor impurities, therefore, brings solar silicon to strong compensation.

This condition requires particular attention for three main reasons:

- At first, because of the relatively high liquid–solid segregation coefficients of B and P ($k_B = 0.8$, $k_P = 0.35$), their concentration cannot be significantly reduced during the solidification of the ingot.
- Then, their presence changes the concentrations of free charge carriers and consequently the specific resistivity and the type (n or p) of the Si-material.[*]
- Eventually, the co-presence of high concentrations of dopants can affect the electronic properties of the material, like lifetime and mobility.

The degree of compensation is usually defined as the compensation ratio RC, that is the ratio between the sum of acceptor (N_A) and donor species (N_D) over their difference, $RC = (N_A + N_D)/(N_A - N_D)$. The different segregation coefficients of B and P would then result in a concentration profile and resistivity distribution along the ingot, and eventually also a carrier-type transition at a certain ingot position, as shown in Figure 1.13 for different values of the compensation ratio.

Eventually, when the dopant concentration in p-type silicon is higher than 10^{17} cm^{-3} not all boron atoms are ionized and a fraction that can reach the 25% is present as non-ionized dopants.

Research work on compensation started in the 1980s [114] but only in the last ten years did interest rapidly grow, in coincidence with the renewed interest of the PV industry in low-cost silicon routes.

---

[*] If the specific resistivity does not stay within a suitable range, a strong negative impact on solar cell efficiency should be expected.

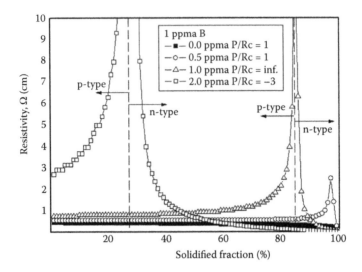

FIGURE 1.13 Resistivity profile along the ingot for different values of the compensation ratio. (Reproduced with permission of CRC Press after M. Tucci et al., Silicon-based photovoltaics, in *Handbook of Silicon Photonics*, L. Vivien and L. Pavesi, Eds., CRC Press, 2013, p. 781–811.)

Different studies have been published on majority carrier mobility characterization in compensated silicon, based on techniques such as the Hall effect [115–121], free carrier absorption [108], electrochemical capacitance voltage [119,122,123] or Fe-acceptor pairing [124]. They indicate a reduction of the majority carrier mobility with the increase of compensation ratio [111–114,125].

The Hall majority carrier mobility of p-type compensated multicrystalline SoG silicon wafers for solar cells was investigated in the temperature range 70–373 K [121]. It was shown that in the range of interest for silicon solar cells (above room temperature) the measured mobility value for a sample with low compensation ratio and low doping density is comparable to the uncompensated references. With decreasing temperatures below approximately 150 K, the difference between samples with low RC and samples with high RC becomes higher.

Furthermore, experimental and modeling studies of Macdonald and Cuevas [126] showed that dopant compensation can also lead to a significant increase of the carrier lifetime in silicon wafers for Auger, radiative, and SRH recombination processes, in good agreement with experimental results [125].

In any case, as the same authors claim, this cannot be considered a general result but occurs when there is an injection dependence of the carrier lifetime. In this case, a positive effect on device parameters occurs as the reduction of mobility is compensated by the increase of lifetime.

It has also been shown [127,128] that the addition of gallium as a dopant to a highly compensated feedstock can help in controlling the resistivity uniformity along the ingot, avoiding the type inversion. By adding gallium, high phosphorus concentrations (4 ppmw) may be tolerated and a cell efficiency close to 17.5% can be reached, in spite of the low carrier mobility.

Several other research groups reported solar cell performance on compensated material comparable with uncompensated control cells [115,129], but that the final result does not depend only on compensation ratio.

The role of doping and of other types of defects that can hamper the lifetime (oxygen aggregates, dislocations, and impurities) should be, in fact, considered and suitably managed. For this reason, n-type compensated silicon solar cells seem more promising, because the LID effect degradation of cell efficiency of compensated n-type Si is less than in p-type CZ-Si.

To understand this last effect better, it was initially proposed that the LID degradation in compensated silicon (see Section 1.4.4.1) depends on the net doping, due to the presence of boron–phosphorus pairs [130,131] but this hypothesis was shown to be inconsistent with a number of other results.

Macdonald et al. [132] assumed that the Voronkov–Falster model for the LID based on B_iO_2 complexes rather than B_sO_2 complexes could be applied also for boron-compensated n-type silicon. Recently, Forster et al. [133] demonstrated that the B–O defect density is systematically found proportional to the total boron concentration, rather than on net doping, showing that compensation cannot reduce LID.

In spite of these recent advancements, it seems, however, that a full understanding of the degradation process in compensated materials has not yet been achieved, leading to an interest in further studies.

1.5 INTEGRAL APPROACH TO MINIMIZE IMPURITIES AND DEFECTS THROUGHOUT THE VALUE CHAIN

The knowledge on how impurities and defects degrade solar cell performance came hand in hand with the research on strategies to minimize their impact, taking advantage also of those adopted by microelectronics. Their application is not straightforward, though, as the quality requirements for microelectronics are very different from those of PV. In particular, microelectronic devices typically sit in a very thin sub-surface region of the silicon wafer while the active area in a solar cell is the whole wafer. Also, in microelectronics the starting point to implement these strategies is typically the availability of high-quality defect-free wafers, while in PV the cheapest possible wafers produced with solar silicon are welcome, even if defective and of overall low quality, provided they do not significantly degrade cell performance.

This implies that the "responsibility" of having a device with enough bulk quality is spread through the value chain, as indicated in Section 1.2, so that an integral strategy should be developed, from the purification steps of the original feedstock to solar cell processing. A schematic representation of this "integral approach" is presented in Figure 1.14.

In general terms, the principles of the strategy, which should be addressed to minimize the impact of impurities and defects, are: first, avoid their incorporation and creation in any step of the Si value chain; and second, in the case this is not possible, reduce their impact by implementing procedures that reduce their electrical activity.

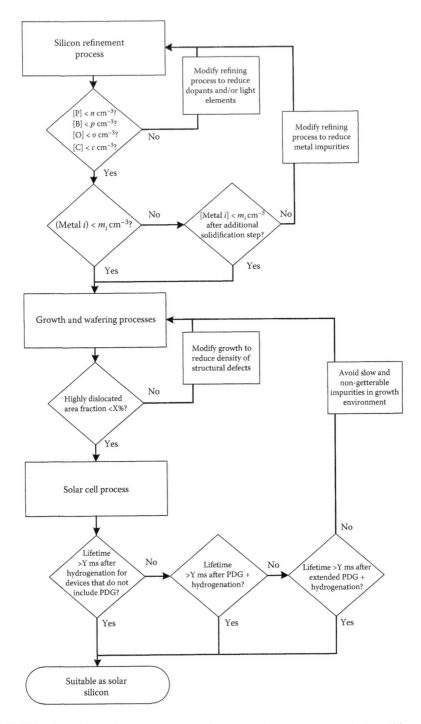

FIGURE 1.14 Schematic representation of an integral approach to track impurities and defects throughout the solar silicon value chain.

The first step in the integral approach has to do with the refinement processes used to produce the solar silicon. Thresholds for the acceptable concentration of impurities should be established, depending on their category: dopants, light elements (C, O, N, …), and metals.

The values n, p, o, c in Figure 1.14 will depend on the performance target that is set for a particular cell technology, and should take into account the evolution these impurities can experience in the subsequent crystallization and solar cell processing steps. The purification process should be modified and improved in case the concentration of dopants and light elements surpasses their thresholds.

Metal impurities in concentrations over the established limits can also make it advisable to revisit the process, or alternatively to add a DS step, that can be very effective in their reduction thanks to their preferential segregation to the liquid.

The establishment of these specifications is especially relevant for the "dirtier" purification processes that want to compete with the incumbent Siemens process. However, as process conditions are sometimes relaxed compared to those of microelectronics, and alternative deposition reactors are tested, the fulfillment of the purity requirements should be also confirmed in the chlorosilane route (see Chapter 5).

Crystallization and wafering steps should pay special attention to the density and distribution of structural defects (grain boundaries and dislocations), so that the areal fraction of highly defective regions in the wafer is kept below a certain threshold $X\%$, otherwise ways to diminish them in the course of the growth step should be investigated.

The key aspects to keep the dislocation density to a minimum are spatial homogeneity in the solidification front and temperature evolution in the process, in particular during the cooling down. A high-temperature post-growth annealing can annihilate dislocations to a certain extent, although side effects can limit the effectiveness of this approach, depending on the material [134–136].

Several advanced casting technologies have been developed recently to achieve materials with a low density of structural defects. This is the case, for instance, of the noncontact crucible (NOC) method, the mono-like casting, or the high-performance multicrystalline (HPMC) growth.

NOC prevents the generation of stress due to the expansion of Si upon cooling by using a seed crystal that induces nucleation on the surface of the Si melt, so that the crystal grows inside the melt without contact with the crucible wall [137]. The dominant defect type in this material is a swirl-defect, likely caused by thermal gradients. A low overall dislocation density $<10^3$ cm^{-2} has been observed in the material, leaving potential to achieve high-efficiency devices [138].

Mono-like casting uses thick CZ wafers as seeds on the bottom of the crucible to control grain size and orientation and to reduce recombination activity at the grain boundaries, with the additional advantage of having large monocrystalline regions that can be more effectively textured [139]. Unfortunately, dislocations, subgrain boundaries, and stacking faults are formed at the seed interfaces, propagating toward the top of the ingot [140], so that the reduction in grain boundary recombination is compensated by the intra-grain recombination. Several solutions are explored today to reduce the regions of high dislocation densities [141].

Fujiwara et al. [142] developed a variant in the casting process based on controlled rapid cooling of the crucible bottom in the start of solidification to promote large grains with controlled orientation. They found that dendrite growth along the bottom wall of the crucible in the initial stage of directional growth is useful for obtaining a polycrystalline structure with large oriented grains. It has been shown that these large grained ingots have fewer dislocations and better electrical properties than the standard mc Si.

This conclusion is just the opposite of that suggested by Yang et al. [143], who found that small initial grains and a high percentage of non-coherent grain boundaries seem to be beneficial to stress relaxation during ingot growth. The initial presence of small grains of random orientation was found not so disadvantageous also considering the final low dislocation density. This variant of the solidification process is now commonly known as HPMC silicon. With this HPMC growth process, the average solar cell efficiency was increased to more than 17.4% using an industrial production cell line.

It is also shown that suitably managing the cooling rate and the axial thermal gradient while keeping the growth rate low, it is possible to increase the grain size. However, in practice, to increase the throughput a higher growth rate is preferred, while a higher thermal gradient is needed to avoid the carbon precipitation. Therefore, a trade-off is to use a higher thermal gradient but using a low growth rate at the early nucleation stage. So, as shown in Reference 35, with a good control of nucleation and grain competition by increasing the undercooling through enhanced uniform spot cooling, grains with more $\Sigma3$- or twin-boundaries were obtained.

The evolution of dislocation clusters in HPMC Si was studied by different techniques in Reference 144, and it was concluded that the termination of dislocation clusters during growth by the interaction with random angle grain boundaries is one of the fundamental processes to reduce the dislocation density.

The crystallization and wafering steps should also be assessed in terms of impurity incorporation. As already said, a very efficient removal mechanism is the preferential segregation of many metal impurities to the Si melt during the ingot growth. Besides, a relevant reduction of impurities can be obtained by an appropriate selection of crucibles and lining coatings [145].

For unconventional crystallization and wafering techniques (ribbon and epi) the same principles apply: the goal is to have a small fraction of highly defective areas and low impurity content. The critical challenges depend largely on the specific conditions at which these processes are conducted: temperature profiles, potential contamination sources in the furnace, etc.

In any case, as it is physically and chemically unavoidable to have a certain level of impurities and defects in the wafer, the key issue would be that their concentration and distribution is such to allow some manipulation and neutralization during the solar cell processing. These strategies are gathered under the concepts "gettering" and "defect engineering," adopted from microelectronics, referring to procedures able to remove harmful impurities from the device (the former), or to reduce their electrical activity (the latter).

An excellent review on gettering and defect engineering can be found in Reference 146, and references therein. The article covers issues concerning the physical properties of transition metals at high temperatures; the classification of the different gettering mechanisms as segregation, relaxation, and injection ones; the description and modeling of those more relevant for PVs (aluminum, phosphorus, and boron gettering); as well as the knowledge about the impurity precipitation and decoration of dislocations and grain boundaries.

As shown in Figure 1.15, gettering can be performed as a three-step process, which proceeds at a sufficiently high temperature. The impurities are first extracted from their equilibrium position (substitutional, precipitated, or trapped at structural defects), into a mobile position that is usually interstitial. Then they move, commonly by diffusion, toward the sink layer, where they are finally trapped.

Among the gettering techniques available for solar cells, phosphorous gettering plays a prominent role, as it is extensively used in standard solar cells associated with the diffusion of the n-type emitter [147–150].

P-diffusion gettering (PDG) efficiency can be limited by the extraction of certain impurities from their fixed position and by the low diffusivity of certain impurities. Several complementary thermal treatments have been explored to enhance the gettering yield in these cases, receiving the name of "extended" PDG [151–158]. These processes have two different goals: first, maintain the wafer at a temperature that is high enough to dissolve precipitates, facilitating the metal extraction and allowing their diffusion to the sink layer; and second, cooling at a slow rate so that the impurities that could not be gettered at least can diffuse to existing precipitates, where they are less harmful.

A defect engineering approach extensively used in solar cell processing is hydrogenation, typically performed thanks to the incorporation of atomic hydrogen from the hydrogenated silicon nitride (SiN$_x$:H) antireflection coating and its diffusion into the bulk during the contact firing step. Hydrogen is said to "passivate" the defects (grain boundaries, dislocations, and impurities) by binding to them and shifting their mid-gap levels to shallow ones, thus reducing their recombination activity.

The process is of course sensitive to SiN$_x$ deposition and post-deposition annealing conditions, and also to the structural defects in the material, as they can influence the effective diffusivity of hydrogen [159–162].

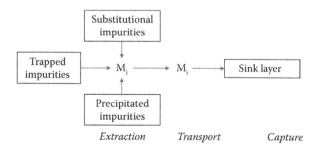

FIGURE 1.15 Conceptual diagram for gettering processes.

It has to be noted that hydrogen passivation can change the "recombination map" resulting from a previous gettering step, especially at the grain boundaries and at dislocations, an aspect that should be taken into account for a precise evaluation of the substrate quality at the end of the solar cell processing [101,163,164].

As Figure 1.14 suggests, the control, evolution, and minimization of impurities is a challenging task, since thermal steps should be appropriately tailored along the value chain, a duty that can be very expensive and time consuming. This tailoring is eased by process simulators that are based on comprehensive models of impurity behavior [68,165–171]. Although they are sometimes limited to certain impurities (typically, Fe) and processes, they can offer guidelines in process optimization. A good cooperative work is presented in Reference 172, which brings together some of these models for Fe, and offers a framework to understand its evolution during crystallization and solar cell processing, distinguishing between a solubility-limited and a diffusivity-limited regime.

1.6 SOLAR SILICON DEFINITION

Some modeling works, founded in the experimental knowledge about impurities and defects that have been reviewed in the previous sections, have been presented to track back the allowable contamination level in the Si feedstock for specific impurities, substrate types, and cell efficiency targets.

The scenario presented in Reference 173, for example, aims at a 15% solar cell on monocrystalline material, and goes back from there in the value chain to show the maximum acceptable contamination level at the cell, wafer, and feedstock levels (see Table 1.3). The liquid–solid segregation effect during crystal growth is able to

TABLE 1.3

Maximum Concentration of Typical Impurities in the Three Stages of Cell, Wafer, and Feedstock Material, for Monocrystalline Si

Impurity	Concentration in Cell (ppma)	Concentration in Wafer (ppma)	Concentration in Feedstock (ppma)
Ti	9.8×10^{-7}	4.8×10^{-7}	1.8×10^{-4}
V	9.4×10^{-6}	5.2×10^{-6}	2.0×10^{-3}
Cr	3.6×10^{-6}	1.5×10^{-5}	0.18
Mn	9.6×10^{-6}	4.0×10^{-5}	0.54
Fe	1.3×10^{-4}	5.6×10^{-4}	9.4
Co	3.0×10^{-5}	1.2×10^{-4}	2.0
Ni	8.4×10^{-4}	3.4×10^{-3}	58
Cu	5.8×10^{-3}	2.4×10^{-2}	8.4
Zn	1.7×10^{-5}	7.0×10^{-5}	0.94
Pd	1.6×10^{-5}	2.6×10^{-4}	36
Pt	7.2×10^{-6}	1.3×10^{-5}	0.07
Au	3.8×10^{-4}	7.0×10^{-4}	38

TABLE 1.4

Acceptable Contamination Levels of Metal Impurities with Distinct Capture Cross Sections and Diffusion Coefficients in mc-Si

Impurity	Concentration in Cell (ppma)	Concentration in Wafer (ppma)	Concentration in Feedstock (ppma)
Ti	2.7×10^{-4}	2.7×10^{-4}	0.022
Cr	4.7×10^{-4}	4.8×10^{-4}	0.026
Fe	9.7×10^{-3}	0.010	12.5
Cu	5.9×10^{-3}	0.046	4.6

Source: Adapted from J. Hofstetter et al., *Mater. Sci. Eng. B*, 159–160, 299–304, 2009.

reduce the impurity concentrations of three or even four orders of magnitude, while the gettering efficacy is lower and depends on the impurity diffusivity.

A similar analysis (see Table 1.4) is carried out in Reference 174, in this case for multicrystalline solar cells and for four representative impurities: titanium (harmful and slow), chromium (harmful and intermediate velocity), iron (intermediate harmfulness and fast), and copper (benign and very fast). Because of the presence of precipitates in mc-Si, a higher total impurity concentration is allowed in mc-Si cells in comparison to the monocrystalline case. The model shows that Fe and Cu may be present in the silicon feedstock in amounts of several ppma, whereas Cr and Ti may only have concentrations of some hundredths ppma.

Coletti [175] compares the impact of some relevant metal impurities for two mc-Si solar cell architectures, one "state-of-the-art in 2010" and the other an "advanced" device architecture. These are defined by some key parameters (thickness, emitter sheet resistance, surface recombination velocity, and rear reflectivity), and achieve 17.5% and 23% solar cell efficiency, respectively. The influence of wafer thickness is also explored.

This model tracks back the impurity concentration from the cell to the feedstock step, assuming the concentration of electrically active impurities is a fraction of the total concentration, calculated from experimental data. As can be seen in Figure 1.16, the impurity with the highest impact is Ti, followed by Cu, Cr, Ni, and Fe, which together form a group two orders of magnitude less sensitive than the former. Advanced devices will be more sensitive to the impurity content than the state-of-the-art cell design, an effect that can be partly compensated by reducing the wafer thickness.

Relevant discrepancies can be found when comparing these model results, as can be observed in the examples above for the case of Cr. A careful look to the characteristics of the scenarios drawn by the authors should be taken into account, both in terms of cell efficiency target and in parameter selection. In addition, it should be noted that these models inevitably simplify the problem to make it manageable, describing with lumped parameters the case of impurities which are present with different charge states, assuming a homogeneous distribution of structural defects

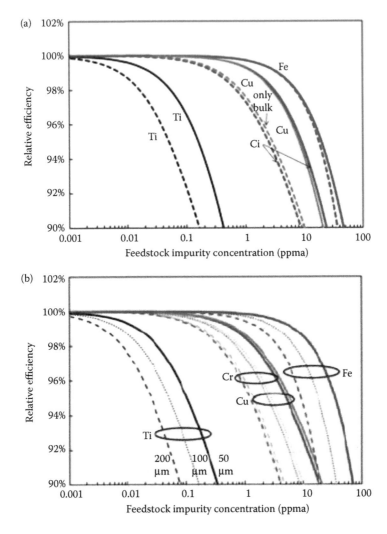

FIGURE 1.16 Relative degradation curve at 90% of a mc-Si ingot versus impurity concentration in the feedstock. (a) State-of-the-art solar cell with 200 μm thickness (full lines) and 100 μm (dashed lines). (b) Advanced cell architecture for three different thicknesses (shown in the figure). (G. Coletti: Sensitivity of state-of-the-art and high efficiency crystalline silicon solar cells to metal impurities. *Prog. Photovolt. Res. Appl.* 2013, 21(5), 1163–1170. Copyright Wiley-VCH Verlag GmbH & Co. KGaA. Reproduced with permission.)

or disregarding aspects such as the potential interaction between different metal impurities.

In any case, these models become a powerful tool, when combined with experimental results, to give some guidance in setting quality targets and feed the discussions that have been held in several technical meetings to standardize a definition of solar silicon.

In 2008 within the European Integrated Project "CrystalClear" a workshop titled "Arriving to Solar Grade (SoG) Silicon Feedstock Specification" was organized in Amsterdam. Different categories for solar silicon were proposed, according to the solar cell performance that could be expected for each of them (see Table 1.5).

The classification, which does not include the high-quality conventional polysilicon produced by the Siemens process, mainly distinguishes by different maximum acceptable levels for B and P:

- Undoped silicon with low B and P concentrations, which will not need significant changes in the crystallization and solar cell processes
- Compensated silicon with higher B and P levels, where process adaptation may be needed

The possibility of using heavily compensated silicon with much higher dopant concentrations was also considered, but threshold values for the different impurities were not established, as modified crystallization and solar cell processes should be developed to make it viable.

A similar approach was followed years later by SEMI, which established a SEMI Standards PV Silicon Materials Task Force for the standardization of solar silicon specifications, responsible for the *Specification for Virgin Silicon Feedstock Materials for Photovoltaic Applications* [176], first published in 2011. The standard distinguishes four categories, as can be seen in Table 1.6.

Category I describes polysilicon purified by the conventional chlorosilane route, while categories II, III, and IV can be related to the undoped, compensated, and heavily compensated silicon of the CrystalClear workshop.

TABLE 1.5

SoG Silicon Feedstock Categories Proposed in the CrystalClear Workshop Held in Amsterdam in 2008

	Undoped Si (ppmw)	Compensated Si (ppmw)
B	<0.05	<0.45
P	<0.1	<0.6
Al	<0.05	<5
Fe	<0.05	<5
Cu	<0.01	<1
Ni	<0.01	<1
Cr	<0.05	<1
Ti	<0.005	<0.05
C	<5	<1 (Cz); <30 (multi)
O	<5	<20

Source: Adapted from G. Coletti, D. Macdonald, and D. Yang, 2012. *Advanced Silicon Materials for Photovoltaic Applications*, pp. 79–125.

TABLE 1.6
SEMI Specification for Solar Silicon Materials

	Category I	Category II	Category III	Category IV
Acceptor concentration (ppba)	<1	<20	<300	<1000
Donor concentration (ppba)	<1	<20	<50	<720
Carbon concentration (ppma)	<0.3	<2	<5	<100
Transition-metal concentration (ppba)	<10	<50	<100	<200
Alkali and alkali-earth metals (ppba)	<10	<50	<100	<4000

Note: Note that in this case the concentrations are in ppb and ppm atomic, while in Table 1.5 they are in ppm weight.

In summary, it should be clear by now that defining the purity requirements for solar silicon is by no means straightforward, and it cannot be done without referring to a specific technology path along the whole Si value chain. In a way, microelectronics faces the same situation but solves it by producing an all-purpose ultrapure EG silicon, purer than needed for some applications, that is affordable because the polysilicon cost is not relevant to the overall device cost, as already mentioned. In PV, on the other hand, it makes sense to divide the feedstock market in segments, each one with a quality target consistent with the processing steps required for a specific cell technology. It cannot be excluded, however, that with the progressive reduction of cost of polysilicon produced with advanced gas-phase processes (see Chapter 5) a low-cost ultrapure silicon will become available also for PV applications.

REFERENCES

1. W. Ballhaus, A. Pagella, C. Vogel, and C. Wilmsmeier, 2012. *Faster, Greener, Smarter—Reaching Beyond the Horizon in the World of Semiconductors*, Price Waterhouse Coopers, Germany.
2. C. del Cañizo, G. del Coso, and W. C. Sinke, 2009. Crystalline silicon solar module technology: Towards the 1 € per watt-peak goal, *Prog. Photovolt. Res. Appl.*, 17(3), 199–209.
3. D. M. Powell, M. T. Winkler, A. Goodrich, and T. Buonassisi, 2013. Modeling the cost and minimum sustainable price of crystalline silicon photovoltaic manufacturing in the United States, *IEEE J. Photovolt.*, 3(2), 662–668.
4. A. Goodrich, P. Hacke, Q. Wang, B. Sopori, R. Margolis, T. L. James, and M. Woodhouse, 2013. A wafer-based monocrystalline silicon photovoltaics road map: Utilizing known technology improvement opportunities for further reductions in manufacturing costs, *Sol. Energy Mater. Sol. Cells*, 114, 110–135.
5. J. R. McCormick, 1985. Polycrystalline silicon technology requirements for photovoltaic applications, in *Silicon Processing for Photovoltaics I*, C. P. Khattak and K. V. Ravi, Eds. Elsevier Science Publishers, North Holland, pp. 1–46.
6. J. Dietl, 1987. Metallurgical ways of silicon meltstock processing, in *Silicon Processing for Photovoltaics II*, C. P. Khattak and K. V. Ravi, Eds. Elsevier Science Publishers, North Holland, pp. 285–351.
7. L. Rogers, 1990. Polysilicon preparation, in *Handbook of Semiconductor Silicon Technology*, Noyes Publications, New Jersey, pp. 33–93.

8. H. A. Aulich and F.-W. Schulze, 2002. Crystalline silicon feedstock for solar cells, *Prog. Photovolt. Res. Appl.*, 10(2), 141–147.

9. E. Schindlbeck and A. Böhm, 2011. Polysilicon—Key to the success of solar energy, in *26th European Photovoltaic Solar Energy Conference*, Hamburg, Germany, pp. 906–908.

10. T. F. Ciszek, 2014. Photovoltaic materials and crystal growth research and development in the Gigawatt era, *J. Cryst. Growth*, 393, 2–6.

11. G. Bye and B. Ceccaroli, 2014. Solar grade silicon: Technology status and industrial trends, *Sol. Energy Mater. Sol. Cells*, 130, 634–646.

12. G. Coletti, D. Macdonald, and D. Yang, 2012. Role of impurities in solar silicon, in *Advanced Silicon Materials for Photovoltaic Applications*, S. Pizzini, Ed., John Wiley & Sons, Chichester, pp. 79–125.

13. S. Pingel, O. Frank, M. Winkler, S. Daryan, T. Geipel, H. Hoehne, and J. Berghold, 2010. Potential induced degradation of solar cells and panels, in *35th IEEE Photovoltaic Specialists Conference*, Honolulu, Hawaii, pp. 002817–002822.

14. P. Hacke, K. Terwilliger, S. Glick, D. Trudell, N. Bosco, S. Johnston, and S. Kurtz, 2010. Test-to-failure of crystalline silicon modules, in *35th IEEE Photovoltaic Specialist Conference*, Honolulu, Hawaii, pp. 244–250.

15. B. Ceccaroli and S. Pizzini, 2012. Processes, in *Advanced Silicon Materials for Photovoltaic Applications*, S. Pizzini, Ed., John Wiley & Sons, Chichester, pp. 21–78.

16. B. Ceccaroli and O. Lohne, 2011. Solar grade silicon feedstock, in *Handbook of Photovoltaic Science and Engineering*, 2nd ed., A. Luque and S. Hegedus, Eds., John Wiley & Sons, Chichester, pp. 169–216.

17. S. Pizzini, 2015. *Physical Chemistry of Semiconductor Materials and Processes*. John Wiley & Sons, Chichester.

18. J. Vedde, T. Clausen, and L. Jensen, 2003. Float-zone silicon for high volume production of solar cells, in *Proceedings of the 3rd World Conference on Photovoltaic Energy Conversion*, Osaka, Japan, pp. 943–946.

19. A. Borghesi, B. Pivac, A. Sassella, and A. Stella, 1995. Oxygen precipitation in silicon growth law for disk precipitates, and oxygen precipitation in silicon effects of ambients on oxygen precipitation in silicon applied physics reviews oxygen precipitation in silicon, *J. Appl. Phys. Lett.*, 77(30), 4169–561.

20. V. V. Voronkov, 1982. The mechanism of swirl defects formation in silicon, *J. Cryst. Growth*, 59(3), 625–643.

21. V. V Voronkov and R. Falster, 1999. Grown-in microdefects, residual vacancies and oxygen precipitation bands in Czochralski silicon, *J. Cryst. Growth*, 204, 462–474.

22. T. Abe, 2000. The formation mechanism of grown-in defects in CZ silicon crystals based on thermal gradients measured by thermocouples near growth interfaces, *Mater. Sci. Eng.* B73, 16–29.

23. T. Abe and T. Takahashi, 2011. Intrinsic point defect behavior in silicon crystals during growth from the melt: A model derived from experimental results, *J. Cryst. Growth*, 334, 16–36.

24. V. V. Voronkov and R. Falster, 1999. Vacancy and self-interstitial concentration incorporated into growing silicon crystals, *J. Appl. Phys.*, 86, 5975.

25. V. Voronkov and R. Falster, 2011. Intrinsic point defects in silicon crystal growth, *Solid State Phen.*, 178–179, 3–14.

26. R. Falster, V. Voronkov, and F. Quast, 2000. On the properties of the intrinsic point defects in silicon: A perspective from crystal growth and wafer processing, *Phys. Status Solidi*, 222, 219–244.

27. E. Dornberger, W. von Ammon, J. Virbulis, B. Hanna, and T. R. Sinno, 2001. Modeling of transient point defect dynamics in Czochralski silicon crystals, *J. Cryst. Growth*, 230, 291–299.

28. V. V. Voronkov and R. Falster, 1998. Vacancy-type microdefect formation in Czochralski silicon, *J. Cryst. Growth*, 194(1), 76–88.

29. N. E. Grant, F. E. Rougieux, D. Macdonald, J. Bullock, and Y. Wan, 2015. Grown-in defects limiting the bulk lifetime of p-type float-zone silicon wafers, *J. Appl. Phys.*, 117(70), 055711.

30. S. Binetti, M. Acciarri, C. Savigni, A. Brianza, S. Pizzini, and A. Musinu, 1996. Effect of nitrogen contamination by crucible encapsulation on polycrystalline silicon material quality, *Mater. Sci. Eng. B*, 36(1–3), 68–72.

31. H. J. Möller, C. Funke, M. Rinio, and S. Scholz, 2005. Multicrystalline silicon for solar cells, *Thin Solid Films*, 487, 79–187.

32. D. Macdonald, A. Cuevas, A. Kinomura, Y. Nakano, and L. J. Geerligs, 2005. Transition-metal profiles in a multicrystalline silicon ingot, *J. Appl. Phys.*, 97(3), 033523.

33. S. Pizzini, M. Acciarri, and S. Binetti, 2005. From electronic grade to solar grade silicon: Chances and challenges in photovoltaics, *Phys. Status Solidi*, 202(15), 2928–2942.

34. T. U. Nærland, L. Arnberg, and A. Holt, 2009. Origin of the low carrier lifetime edge zone in multicrystalline PV silicon, *Prog. Photovolt. Res. Appl.*, 17, 289–296.

35. C. W. Lan, W. C. Lan, T. F. Lee, A. Yu, Y. M. Yang, W. C. Hsu, B. Hsu, and A. Yang, 2012. Grain control in directional solidification of photovoltaic silicon, *J. Cryst. Growth*, 360(1), 68–75.

36. J. P. Kalejs, 2004. Silicon ribbons for solar cells, *Solid State Phen.*, 95–96, 159–174.

37. J. I. Hanoka, 2001. An overview of silicon ribbon growth technology, *Sol. Energy Mater. Sol. Cells*, 65(1–4), 231–237.

38. G. Hahn and P. Geiger, 2003. Record efficiencies for EFG and string ribbon solar cells, *Prog. Photovolt. Res. Appl.*, 11(5), 341–346.

39. S. Binetti, S. Ferrari, M. Acciarri, S. Acerboni, R. Canteri, and S. Pizzini, 1993. New evidences about carbon and oxygen segregation processes in polycrystalline silicon, *Solid State Phen.*, 32–33, 181–190.

40. R. Slunjski, I. Capan, B. Pivac, A. Le Donne, and S. Binetti, 2011. Effects of low-temperature annealing on polycrystalline silicon for solar cells, *Sol. Energy Mater. Sol. Cells*, 95(2), 559–563.

41. T. Buonassisi et al. 2006. Chemical natures and distributions of metal impurities in multicrystalline silicon materials, *Prog. Photovolt. Res. Appl.*, 14(6), 512–531.

42. H. Rodriguez, I. Guerrero, W. Koch, A. L. Endrös, D. Franke, C. Hässler, J. P. Kalejs, and H. J. Möller, 2011. Bulk crystal growth and wafering for PV, in *Handbook of Photovoltaic Science and Engineering*, A. Luque and S. Hegedus, Eds., John Wiley & Sons, Chichester, pp. 218–264.

43. W. Schockley and H. J. Queisser, 1961. Detailed balance limit of efficiency of P-N junction solar cells, *J. Appl. Phys.*, 32(3), 510–519.

44. C. H. Henry, 1980. Limiting efficiencies of ideal single and multiple energy gap terrestrial solar cells, *J. Appl. Phys.*, 51(8), 4494–4500.

45. T. Tiedje, E. Yablonovitch, G. D. Cody, and B. G. Brooks, 1984. Limiting efficiency of silicon solar cells, *IEEE Trans. Electron Devices*, 31(5), 711–716.

46. A. Richter, M. Hermle, and S. W. Glunz, 2013. Reassessment of the limiting efficiency for crystalline silicon solar cells, *IEEE J. Photovolt.*, 3(4), 1184–1191.

47. R. Swanson, 2005. Approaching the 29% limit efficiency of silicon solar cells, in *31st IEEE Photovoltaic Specialist Conference*, Lake Buena Vista, Florida, pp. 889–894.

48. S. W. Glunz, R. Preu, and D. Biro, 2012. Crystalline silicon solar cells: State-of-the-art and future developments, in *Comprehensive Renewable Energy. Vol. 1: Photovoltaic Solar Energy*, A. Sayigh, Ed., Elsevier, Amsterdam, pp. 353–387.

49. H. Wagner, T. Ohrdes, A. Dastgheib-Shirazi, B. Puthen-Veettil, D. König, D. Konig, and P. P. Altermatt, 2014. A numerical simulation study of gallium-phosphide/silicon heterojunction passivated emitter and rear solar cells, *J. Appl. Phys.*, 115(4), 44506–44508.

50. H. Wagner, J. Hofstetter, B. Mitchell, P. P. Altermatt, and T. Buonassisi, 2015. Device architecture and lifetime requirements for high efficiency multicrystalline silicon solar cells, *Energy Procedia*, 77, 225–230.
51. J. Hofstetter, C. del Cañizo, H. Wagner, S. Castellanos, and T. Buonassisi, 2016. Materials requirements for the adoption of unconventional crystal-growth techniques for high-efficiency solar cells, *Prog. Photovolt. Res. Appl.*, 24, 122–132.
52. D. Macdonald and A. Cuevas, 2000. Reduced fill factors in multicrystalline silicon solar cells due to injection-level dependent bulk recombination lifetimes, *Prog. Photovolt. Res. Appl.*, 8, 363–375.
53. B. Michl, M. Kasemann, W. Warta, and M. C. Schubert, 2013. Wafer thickness optimization for silicon solar cells of heterogeneous material quality, *Phys. Status Solidi—Rapid Res. Lett.*, 7(11), 955–958.
54. G. E. Wakefield, P. D. Maycock, and T. L. Chu, 1975. Influence of impurities in silicon solar cell performance, in *11th IEEE Photovoltaic Specialist Conference*, Scottsdale, Arizona, pp. 49–55.
55. J. R. Davis, A. Rohatgi, R. H. Hopkins, P. D. Blais, P. Rai-Choudhury, J. R. McCormick, and H. C. Mollenkopf, 1980. Impurities in silicon solar cells, *IEEE Trans. Electron Devices*, 27(4), 677–687.
56. G. Coletti, P. C. P Bronsveld, G. Hahn, W. Warta, D. Macdonald, B. Ceccaroli, K. Wambach, N. Le Quang, and J. M. Fernandez, 2011. Impact of metal contamination in silicon solar cells, *Adv. Funct. Mater.*, 21, 879–890.
57. S. Pearton, 1999. Impurities in silicon, in *Properties of Crystalline Silicon*, R. Hull, Ed., The Institution of Electrical Engineers, London, pp. 477–596.
58. K. Graff, 1995. *Metal Impurities in Silicon-Device Fabrication*. Springer-Verlag, Berlin.
59. E. R. Weber, 1983. Transition metals in silicon, *Appl. Phys. A Solids Surf.*, 30(1), 1–22.
60. D. V. Lang, 1974. Deep-level transient spectroscopy: A new method to characterize traps in semiconductors, *J. Appl. Phys.*, 45, 3023.
61. S. Rein, T. Rehrl, W. Warta, and S. W. Glunz, 2002. Lifetime spectroscopy for defect characterization: Systematic analysis of the possibilities and restrictions, *J. Appl. Phys.*, 91(3), 2059–2070.
62. J. Schmidt, B. Lim, D. Walter, K. Bothe, S. Gatz, T. Dullweber, and P. P. Altermatt, 2013. Impurity-related limitations of next-generation industrial silicon solar cells, *IEEE J. Photovolt.*, 3(1), 114–118.
63. S. Pizzini, L. Bigoni, M. Beghi, and C. Chemelli, 1986. On the effect of impurities on the photovoltaic behavior of solar grade silicon: II. Influence of titanium, vanadium, chromium, iron, and zirconium on photovoltaic behavior of polycrystalline solar cells, *J. Electrochem. Soc.*, 133(11), 2363–2373.
64. S. Pizzini, 1988. Influence of extended defects and native impurities on the electrical properties of directionally solidified polycrystalline silicon, *J. Electrochem. Soc.*, 135(1), 155–165.
65. W. Bergholz and D. Gilles, 2000. Impact of research on defects in silicon on the microelectronic industry, *Phys. Status Solidi*, 222(5), 5–23.
66. M. Seibt and K. Graff, 1988. Characterization of haze-forming precipitates in silicon, *J. Appl. Phys.*, 63(9), 4444–4450.
67. P. S. Plekhanov, R. Gafiteanu, U. M. Gösele, and T. Y. Tan, 1999. Modeling of gettering of precipitated impurities from Si for carrier lifetime improvement in solar cell applications, *J. Appl. Phys.*, 86(5), 2453.
68. C. del Cañizo and A. Luque, 2000. A comprehensive model for the gettering of lifetime-killing impurities in silicon, *J. Electrochem. Soc.*, 147(7), 2685–2692.
69. P. S. Plekhanov and T. Y. Tan, 2000. Schottky effect model of electrical activity of metallic precipitates in silicon, *Appl. Phys. Lett.*, 76(25), 3777–3779.

70. M. D. Negoita and T. Y. Tan, 2003. Metallic precipitate contribution to generation and recombination currents in p-n junction devices due to the Schottky effect, *J. Appl. Phys.*, 94(8), 5064–5070.

71. W. Kwapil, J. Sch, F. Schindler, W. Warta, and M. C. Schubert, 2014. Impact of iron precipitates on carrier lifetime in silicon wafers in model and experiment, *IEEE J. Photovolt.*, 4(3), 791–798.

72. A. A. Istratov, P. Zhang, R. J. McDonald, A. R. Smith, M. Seacrist, J. Moreland, J. Shen, R. Wahlich, and E. R. Weber, 2005. Nickel solubility in intrinsic and doped silicon, *J. Appl. Phys.*, 97(2), 023505.

73. T. Buonassisi et al., 2005. Synchrotron-based investigations of the nature and impact of iron contamination in multicrystalline silicon solar cells, *J. Appl. Phys.*, 97(7), 74901.

74. M. Seibt, R. Khalil, V. Kveder, and W. Schröter, 2009. Electronic states at dislocations and metal silicide precipitates in crystalline silicon and their role in solar cell materials, *Appl. Phys. A Mater. Sci. Process.*, 96, 235–253.

75. C. Donolato, 1998. Modeling the effect of dislocations on the minority carrier diffusion length of a semiconductor, *J. Appl. Phys.*, 84(5), 2656–2664.

76. V. Kveder, M. Kittler, and W. Schröter, 2001. Recombination activity of contaminated dislocations in silicon: A model describing electron-beam-induced current contrast behavior, *Phys. Rev. B*, 63(11), 1–11.

77. M. Rinio, A. Yodyungyong, S. Keipert-Colberg, D. Borchert, and A. Montesdeoca-Santana, 2011. Recombination in ingot cast silicon solar cells, *Phys. Status Solidi*, 208(4), 760–768.

78. G. Stokkan, S. Riepe, O. Lohne, and W. Warta, 2007. Spatially resolved modeling of the combined effect of dislocations and grain boundaries on minority carrier lifetime in multicrystalline silicon, *J. Appl. Phys.*, 101(5), 53515–53519.

79. M. Kittler, W. Seifert, and O. Krüger, 2001. Electrical behaviour of crystal defects in silicon solar cells, *Solid State Phen.*, 78–79, 39–48.

80. S. Kusanagi, T. Sekiguchi, B. Shen, and K. Sumino, 1995. Electrical activity of extended defects and gettering of metallic impurities in silicon, *Mater. Sci. Technol.*, 11, 685–690.

81. B. Shen, T. Sekiguchi, J. Jablonski, and K. Sumino, 1994. Gettering of copper by bulk stacking faults and punched-out dislocations in Czochralski-grown silicon, *J. Appl. Phys.*, 76(8), 4540.

82. S. Acerboni, S. Pizzini, S. Binetti, M. Acciarri, and B. Pichaud, 1994. Effect of oxygen aggregation processes on the recombining activity of 60° dislocations in Czochralski grown silicon, *J. Appl. Phys.*, 76(5), 2703.

83. M. Tajima, M. Ikebe, Y. Ohshita, and A. Ogura, 2010. Photoluminescence analysis of Iron contamination effect in multicrystalline silicon wafers for solar cells, *J. Electron. Mater.*, 39(6), 747–750.

84. N. A. Drozdov, A. A. Patrin, and V. D. Tkachev, 1976. Recombination radiation on dislocations in silicon, *JETP Lett.*, 23(11), 597–599.

85. M. Tajima, H. Takeno, and T. Abe, 1992. Characterization of point-defect in Si crystals by highly spatially resolved photoluminescence, *Mater. Sci. Forum*, 83–87, 1327–1332.

86. V. Kveder, T. Sekiguchi, and K. Sumino, 1995. Electronic states associated with dislocations in p-type silicon studied by means of electric-dipole spin resonance and deep-level transient spectroscopy, *Phys. Rev. B*, 51(23), 16751.

87. S. Binetti, S. Pizzini, E. Leoni, R. Somaschini, A. Castaldini, and A. Cavallini, 2002. Optical properties of oxygen precipitates and dislocations in silicon, *J. Appl. Phys.*, 92(5), 2437.

88. S. Binetti, R. Somaschini, A. Le Donne, E. Leoni, S. Pizzini, D. Li, and D. Yang, 2002. Dislocation luminescence in nitrogen-doped Czochralski and float zone silicon, *J. Phys. Condens. Matter*, 14(14), 13247–13254.

89. M. Tajima, Y. Iwata, F. Okayama, H. Toyota, H. Onodera, and T. Sekiguchi, 2012. Deep-level photoluminescence due to dislocations and oxygen precipitates in multicrystalline Si, *J. Appl. Phys.*, 111(11), 113523.

90. V. Higgs, E. C. Lightowlers, X. Xiaoa, and J. C. Sturm, 1994. Characterization of Si/ $Si_{1-x}Ge_x$/Si quantum wells by cathodoluminescence imaging and spectroscopy, *Appl. Phys. Lett.*, 64(5), 607–609.

91. V. Higgs, E. C. Lightowlers, S. Tajbakhsh, and P. J. Wright, 1992. Cathodoluminescence imaging and spectroscopy of dislocations in Si and $Si_{1-x}Ge_x$ alloys, *Appl. Phys. Lett.*, 61(9), 1087.

92. M. Tajima and H. Sugimoto, 2009. Defect analysis in solar cell silicon by photoluminescence spectroscopy and topography, *ECS Trans.*, 25(3), 3–10.

93. S. Binetti, A. Le Donne, and A. Sassella, 2014. Photoluminescence and infrared spectroscopy for the study of defects in silicon for photovoltaic applications, *Sol. Energy Mater. Sol. Cells*, 130, 696–703.

94. S. Binetti, A. Le Donne, and M. Acciarri, 2005. Processing step-related upgrading of silicon-based solar cells detected by photoluminescence spectroscopy, *Sol. Energy Mater. Sol. Cells*, 86(1), 11–18.

95. H. T. Nguyen, F. E. Rougieux, F. Wang, and D. Macdonald, 2015. Effects of solar cell processing steps on dislocation luminescence in multicrystalline silicon, *Energy Procedia*, 77, 619–625.

96. J. Chen, T. Sekiguchi, D. Yang, F. Yin, K. Kido, and J. Chen, 2004. Electron-beam-induced current study of grain boundaries in multicrystalline silicon, *J. Appl. Phys.*, 96(10), 5490–5495.

97. M. Acciarri, S. Binetti, A. Le Donne, S. Marchionna, M. Vimercati, J. Libal, R. Kopecek, and K. Wambach, 2007. Effect of P-induced gettering on extended defects in n-type multicrystalline silicon, *Prog. Photovoltaics Res. Appl.*, 15(5), 375–386.

98. T. Buonassisi, A. A. Istratov, M. D. Pickett, M. A. Marcus, T. F. Ciszek, and E. R. Weber, 2006. Metal precipitation at grain boundaries in silicon: Dependence on grain boundary character and dislocation decoration, *Appl. Phys. Lett.*, 89(4), 1–4.

99. S. Castellanos, M. Kivambe, J. Hofstetter, M. Rinio, B. Lai, and T. Buonassisi, 2014. Variation of dislocation etch-pit geometry: An indicator of bulk microstructure and recombination activity in multicrystalline silicon, *J. Appl. Phys.*, 115(18), 183511.

100. J. Chen, D. Yang, Z. Xi, and T. Sekiguchi, 2005. Electron-beam-induced current study of hydrogen passivation on grain boundaries in multicrystalline silicon: Influence of GB character and impurity contamination, *Phys. B Condens. Matter*, 364(1–4), 162–169.

101. L. J. Geerligs, Y. Komatsu, I. Röver, K. Wambach, I. Yamaga, and T. Saitoh, 2007. Precipitates and hydrogen passivation at crystal defects in n- and p-type multicrystalline silicon, *J. Appl. Phys.*, 102(9), 093702.

102. J. Bauer, A. Hähnel, P. Werner, N. Zakharov, H. Blumtritt, A. Zuschlag, and O. Breitenstein, 2016. Recombination at Lomer dislocations in multicrystalline silicon for solar cells, *IEEE J. Photovoltaics*, 6(1), 100–110.

103. H. Fischer and W. Pschunder, 1973. Investigation of photon and thermal induced changes in silicon solar cells, in *10th IEEE Photovoltaic Specialist Conference*, Palo Alto, California, pp. 404–411.

104. J. Schmidt, A. G. Aberle, and R. Hezel, 1997. Investigation of carrier lifetime instabilities in Cz-grown silicon, in *26th IEEE Photovoltaic Specialists Conference*, Anaheim, California, pp. 13–18.

105. J. Schmidt, 2004. Light-induced degradation in crystalline silicon solar cells, *Solid State Phenom.*, 95–96, 187–196.

106. J. Schmidt, K. Bothe, and R. Hezel, 2003. Structure and transformation of the metastable centre in Cz-silicon solar cells, *Proceedings of 3rd World Conference on Photovoltaic Energy Conversion, 2003*, 3, 2887–2892.

107. K. Bothe and J. Schmidt, 2006. Electronically activated boron-oxygen-related recombination centers in crystalline silicon, *J. Appl. Phys.*, 99(1), 013701.
108. D. Macdonald et al., 2009. Light-induced boron-oxygen defect generation in compensated p-type Czochralski silicon, *J. Appl. Phys.*, 105, 93704.
109. V. V. Voronkov and R. Falster, 2010. Latent complexes of interstitial boron and oxygen dimers as a reason for degradation of silicon-based solar cells, *J. Appl. Phys.*, 107(5), 53509.
110. A. Herguth, G. Schubert, M. Kaes, G. Hahn, 2008. Investigations on the long time behavior of the metastable boron-oxygen complex in crystalline silicon, *Progr. Photovolt.: Res. Appl.*, 16, 135–140.
111. S. Bernardini, D. Saynova, S. Binetti, and G. Coletti, 2012. Light-induced degradation in compensated mc-Si p-type solar cells, in *38th IEEE Photovoltaic Specialists Conference*, Austin, Texas, pp. 3242–3247.
112. S. K. Estreicher, M. Stavola, and J. Weber, 2014. Hydrogen in Si and Ge, in *Silicon, Germanium, and Their Alloys: Growth, Defects, Impurities, and Nanocrystals*, G. Kissinger and S. Pizzini, Eds. CRC Press, Boca Raton, Florida, pp. 217–254.
113. M. Tucci, M. Izzi, R. Kopecek, M. McCann, A. Le Donne, S. Binetti, S. Huang, and G. Conibeer, 2013. Silicon-based photovoltaics, in *Handbook of Silicon photonics*, L. Vivien and L. Pavesi, Eds. CRC Press, Boca Raton, Florida, p. 781.
114. S. Pizzini and C. Calligarich, 1984. On the effect of impurities on the photovoltaic behavior of solar grade silicon: I. The role of boron and phosphorous primary impurities in p-type single-crystal silicon, *J. Electrochem. Soc.*, 131(9), 2128–2132.
115. J. Libal, S. Novaglia, M. Acciarri, S. Binetti, R. Petres, J. Arumughan, R. Kopecek, and A. Prokopenko, 2008. Effect of compensation and of metallic impurities on the electrical properties of Cz-grown solar grade silicon, *J. Appl. Phys.*, 104(10), 104507.
116. J. Veirman, S. Dubois, N. Enjalbert, J. P. Garandet, D. R. Heslinga, and M. Lemiti, 2010. Hall mobility reduction in single-crystalline silicon gradually compensated by thermal donors activation, *Solid. State. Electron.*, 54(6), 671–674.
117. R. Zierer, G. Gärtner, and H. J. Möller, Determination of dopant concentrations in UMG-silicon by FTIR spectroscopy, Hall- and resistivity measurements, in *25th European Photovoltaic Solar Energy Conference*, 2010, 1318–1321.
118. F. Schindler, J. Geilker, W. Kwapil, J. Giesecke, M. C. Schubert, and W. Warta, Conductivity mobility and Hall mobility in compensated multicrystalline silicon, in *European Photovoltaic Solar Energy Conference*, 2010, pp. 2364–2368.
119. F. E. Rougieux, D. Macdonald, A. Cuevas, S. Ruffell, J. Schmidt, B. Lim, and A. P. Knights, 2010. Electron and hole mobility reduction and Hall factor in phosphorus-compensated p-type silicon, *J. Appl. Phys.*, 108, 13706.
120. J. Veirman, S. Dubois, N. Enjalbert, J. P. Garandet, and M. Lemiti, 2011. Electronic properties of highly-doped and compensated solar-grade silicon wafers and solar cells, *J. Appl. Phys.*, 109(10), 103711.
121. C. A. M. Modanese, S. Binetti, A. K. Søiland, M. Di Sabatino, L. Arnberg, 2013. Temperature-dependent Hall-effect measurements of p-type multicrystalline compensated solar grade silicon, *Prog. Photovolt. Res.* 21(7), 1469–1477.
122. B. Lim, M. Wolf, and J. Schmidt, 2011. Carrier mobilities in multicrystalline silicon wafers made from UMG-Si, *Phys. Status Solidi (c)*, 8(3), 835–838.
123. B. Lim, A. Liu, D. Macdonald, K. Bothe, and J. Schmidt, 2009. Impact of dopant compensation on the deactivation of boron-oxygen recombination centers in crystalline silicon, *Appl. Phys. Lett.*, 95, 232109.
124. D. Macdonald, A. Cuevas, and L. J. Geerligs, 2008. Measuring dopant concentrations in compensated p-type crystalline silicon via iron-acceptor pairing, *Appl. Phys. Lett.*, 92(20), 202119.
125. S. Dubois, N. Enjalbert, and J. P. Garandet, 2008. Effects of the compensation level on the carrier lifetime of crystalline silicon, *Appl. Phys. Lett.*, 93(3), 32114.

126. D. Macdonald and A. Cuevas, 2011. Recombination in compensated crystalline silicon for solar cells, *J. Appl. Phys.*, 109(4), 43704.

127. M. Forster, E. Fourmond, R. Einhaus, H. Lauvray, J. Kraiem, and M. Lemiti, 2011. Ga co-doping in Cz-grown silicon ingots to overcome limitations of B and P compensated silicon feedstock for PV applications, *Phys. Status Solidi*, 8(3), 678–681.

128. A. Cuevas, M. Forster, F. Rougieux, and D. Macdonald, 2012. Compensation engineering for silicon solar cells, *Energy Procedia*, 15, 67–77.

129. S. Rein, W. Kwapil, J. Geilker, G. Emanuel, M. Spitz, I. Reis, A. Weil et al. 2009. Impact of compensated solar-grade silicon on Czochralski silicon wafers and solar cells, in *24th European Photovoltaic Solar Energy Conference*, Hamburg, Germany, pp. 1140–1147.

130. R. Kopecek, J. Arumughan, K. Peter, E. A. Good, J. Libal, M. Acciarri, and S. Binetti, 2008. Crystalline Si solar cells from compensated material: Behaviour of light induced degradation, in *23th European Photovoltaic Solar Energy Conference*, Valencia, Spain, pp. 1855–1858.

131. W. Krühler, C. Moser, F. Schulze, and H. Aulich, 1988. Effect of phosphorus compensation on the electronic properties of solar-grade silicon, in *8th European Photovoltaic Solar Energy Conference*, Florence, Italy, pp. 1181–1185.

132. D. Macdonald, A. Liu, A. Cuevas, B. Lim, and J. Schmidt, 2011. The impact of dopant compensation on the boron-oxygen defect in p- and n-type crystalline silicon, *Phys. Status Solidi*, 208(3), 559–563.

133. M. Forster, P. Wagner, J. Degoulange, R. Einhaus, G. Galbiati, F. E. Rougieux, A. Cuevas, and E. Fourmond, 2014. Impact of compensation on the boron and oxygen-related degradation of upgraded metallurgical-grade silicon solar cells, *Sol. Energy Mater. Sol. Cells*, 120, 390–395.

134. K. Hartman, M. Bertoni, J. Serdy, and T. Buonassisi, 2008. Dislocation density reduction in multicrystalline silicon solar cell material by high temperature annealing, *Appl. Phys. Lett.*, 93(12), 122108.

135. M. Bertoni, C. Colin, and T. Buonassisi, 2009. Dislocation engineering in multicrystalline silicon, *Solid State Phenom.*, 156–158, 11–18.

136. S. Castellanos and T. Buonassisi, 2015. Dislocation density reduction limited by inclusions in kerfless high-performance multicrystalline silicon, *Phys. Status Solidi—Rapid Res. Lett.*, 9(9), 503–506.

137. K. Nakajima, K. Morishita, R. Murai, and K. Kutsukake, 2012. Growth of high-quality multicrystalline Si ingots using noncontact crucible method, *J. Cryst. Growth*, 355(1), 38–45.

138. K. Nakajima, R. Murai, S. Ono, K. Morishita, and M. M. Kivambe, 2015. Shape and quality of Si single bulk crystals grown inside Si melts using the noncontact crucible method, *Jpn. J. Appl. Phys.*, 54, 015504.

139. N. Stoddard, B. Wu, I. Witting, M. C. Wagener, Y. Park, G. A. Rozgonyi, and R. Clark, 2008. Casting single crystal silicon: Novel defect profiles from BP Solar's Mono2 TM wafers, *Solid State Phenom.*, 131–133, 1–8.

140. T. Ervik, G. Stokkan, T. Buonassisi, Ø. Mjøs, and O. Lohne, 2014. Dislocation formation in seeds for quasi-monocrystalline silicon for solar cells, *Acta Mater.*, 67, 199–206.

141. D. Hu, S. Yuan, L. He, H. Chen, Y. Wan, X. Yu, and D. Yang, 2015. Higher quality mono-like cast silicon with induced grain boundaries, *Sol. Energy Mater. Sol. Cells*, 140, 121–125.

142. K. Fujiwara, W. Pan, N. Usami, K. Sawada, M. Tokairin, Y. Nose, A. Nomura, T. Shishido, and K. Nakajima, 2006. Growth of structure-controlled polycrystalline silicon ingots for solar cells by casting, *Acta Mater.*, 54(12), 3191–3197.

143. Y. M. Yang, A. Yu, B. Hsu, W. C. Hsu, A. Yang, and C. W. Lan, 2015. Development of high-performance multicrystalline silicon for photovoltaic industry, *Prog. Photovolt. Res. Appl.*, 23, 340–351.

144. G. Stokkan, Y. Hu, Ø. Mjøs, and M. Juel, 2014. Study of evolution of dislocation clusters in high performance multicrystalline silicon, *Sol. Energy Mater. Sol. Cells*, 130, 679–685.

145. R. Kvande, L. Arnberg, and C. Martin, 2009. Influence of crucible and coating quality on the properties of multicrystalline silicon for solar cells, *J. Cryst. Growth*, 311(3), 765–768.

146. M. Seibt and V. Kveder, 2012. Gettering processes and the role of extended defects, in *Advanced Silicon Materials for Photovoltaic Applications*, S. Pizzini, Ed., John Wiley & Sons, Chichester, 127–188.

147. A. Ourmazd and W. Schröter, 1984. Phosphorus gettering and intrinsic gettering of nickel in silicon, *Appl. Phys. Lett.*, 45(7), 781.

148. J. S. Kang and D. K. Schroder, 1989. Gettering in silicon, *J. Appl. Phys.*, 65(8), 2974–2985.

149. S. A. McHugo, H. Hiesimair, and E. R. Weber, 1997. Gettering of metallic impurities in photovoltaic silicon, *Appl. Phys. A Mater. Sci. Process.*, 64(2), 127–137.

150. S. M. Myers, M. Seibt, and W. Schröter, 2000. Mechanisms of transition-metal gettering in silicon, *J. Appl. Phys.*, 88(7), 3795–3819.

151. J. Härkönen, V.-P. Lempinen, T. Juvonen, and J. Kylmäluoma, 2002. Recovery of minority carrier lifetime in low-cost multicrystalline silicon, *Sol. Energy Mater. Sol. Cells*, 73(2), 125–130.

152. T. Buonassisi, A. A. Istratov, M. A. Marcus, B. Lai, Z. Cai, S. M. Heald, and E. R. Weber, 2005. Engineering metal-impurity nanodefects for low-cost solar cells, *Nat. Mater.*, 4, 676–679.

153. P. Manshanden and L. J. Geerligs, 2006. Improved phosphorous gettering of multicrystalline silicon, *Sol. Energy Mater. Sol. Cells*, 90(7–8), 998–1012.

154. M. D. Pickett and T. Buonassisi, 2008. Iron point defect reduction in multicrystalline silicon solar cells, *Appl. Phys. Lett.*, 92(12), 4–6.

155. M. Rinio, A. Yodyunyong, S. Keipert-Colberg, Y. P. B. Mouafi, D. Borchert, and A. Montesdeoca-Santana, 2011. Improvement of multicrystalline silicon solar cells by a low temperature anneal after emitter diffusion, *Prog. Photovolt. Res. Appl.*, 19(2), 165–169.

156. J. Hofstetter, J. F. Lelièvre, C. del Cañizo, and A. Luque, 2009. Study of internal versus external gettering of iron during slow cooling processes for silicon solar cell fabrication, *Solid State Phen.*, 156–158, 387–393.

157. D. P. Fenning, A. S. Zuschlag, M. I. Bertoni, B. Lai, G. Hahn, and T. Buonassisi, 2013. Improved iron gettering of contaminated multicrystalline silicon by high-temperature phosphorus diffusion, *J. Appl. Phys.*, 113(21), 214504.

158. J. Hofstetter, D. P. Fenning, D. M. Powell, A. E. Morishige, and H. Wagner, 2014. Sorting metrics for customized phosphorus diffusion gettering, *IEEE J. Photovolt.*, 4(6), 1421–1428.

159. O. F. Vyvenko, O. Krüger, and M. Kittler, 2000. Cross-sectional electron-beam-induced current analysis of the passivation of extended defects in cast multicrystalline silicon by remote hydrogen plasma treatment, *Appl. Phys. Lett.*, 76(6), 697–699.

160. S. Martinuzzi, I. Périchaud, and F. Warchol, 2003. Hydrogen passivation of defects in multicrystalline silicon solar cells, *Sol. Energy Mater. Sol. Cells*, 80(3), 343–353.

161. C. E. Dubé and J. I. Hanoka, 2005. Hydrogen passivation of multicrystalline silicon, in *IEEE Photovoltaic Specialist Conference*, pp. 883–888.

162. G. Hahn, M. Käs, and B. Herzog, 2009. Hydrogenation in crystalline silicon materials for photovoltaic application, *Solid State Phenom.*, 156–158, 343–349.

163. S. Gindner, P. Karzel, B. Herzog, and G. Hahn, 2014. Efficacy of phosphorus gettering and hydrogenation in multicrystalline silicon, *IEEE J. Photovolt.*, 4(4), 1–8.

164. S. Pizzini, M. Acciarri, S. Binetti, D. Narducci, and C. Savigni, 1997. Recent achievements in semiconductor defect passivation, *Mater. Sci. Eng. B*, B45(1–3), 126–133.

165. V. V Kveder, W. Schroter, A. Sattler, and M. Seibt, 2000. Simulation of Al and phosphorus diffusion gettering in Si, *Mater. Sci. Eng. B*, 71(1–3), 175–181.
166. H. Hieslmair, S. Balasubramanian, A. A. Istratov, and E. R. Weber, 2001. Gettering simulator: Physical basis and algorithm, *Semicond. Sci. Technol.*, 16(2001), 567–574.
167. A. L. Smith, K. Wada, and L. Kimerling, 1999. Modeling of transition metal redistribution to enable wafer design for gettering, *J. Electrochem. Soc.*, 147(3), 1154–1160.
168. A. Haarahiltunen, H. Savin, M. Yli-Koski, H. Talvitie, and J. Sinkkonen, 2009. Modeling phosphorus diffusion gettering of iron in single crystal silicon, *J. Appl. Phys.*, 105(2), 2–6.
169. J. Schön, H. Habenicht, M. C. Schubert, and W. Warta, 2011. Understanding the distribution of iron in multicrystalline silicon after emitter formation: Theoretical model and experiments, *J. Appl. Phys.*, 109(6), 063717.
170. J. Hofstetter, D. P. Fenning, M. I. Bertoni, J. F. Lelièvre, C. del Cañizo, and T. Buonassisi, 2011. Impurity-to-efficiency simulator: Predictive simulation of silicon solar cell performance based on iron content and distribution, *Prog. Photovolt. Res. Appl.*, 19(4), 487–497.
171. R. Chen, B. Trzynadlowski, and S. T. Dunham, 2014. Phosphorus vacancy cluster model for phosphorus diffusion gettering of metals in Si, *J. Appl. Phys.*, 115(5), 054906.
172. A. Morishige, H. Laine, J. Schön, A. Haarahiltunen, J. Hofstetter, C. del Cañizo, M. Schubert, H. Savin, and T. Buonassisi, 2015. Building intuition of iron evolution during solar cell processing through rigorous analysis of different process models, *Appl. Phys. A Mater. Sci. Process.*, 120, 1357–1373.
173. C. del Cañizo, A. Luque, J. Bullón, Á. Miranda, J. Míguez, H. Riemann, and N. Abrosimov, 2005. Integral procedure for the fabrication of solar cells on solar silicon, in *EUPVSEC*, June, pp. 946–949.
174. J. Hofstetter, J. F. Lelièvre, C. del Cañizo, and A. Luque, 2009. Acceptable contamination levels in solar grade silicon: From feedstock to solar cell, *Mater. Sci. Eng. B*, 159–160, 299–304.
175. G. Coletti, 2013. Sensitivity of state-of-the-art and high efficiency crystalline silicon solar cells to metal impurities, *Prog. Photovolt. Res. Appl.*, 21(5), 1163–1170.
176. SEMI Document PV17-0611, Specification for Virgin Silicon Feedstock Materials for Photovoltaic Applications, 2011.

2 The MG Silicon Route

Eivind Øvrelid and Sergio Pizzini

CONTENTS

2.1 INTRODUCTION

Starting in the late 1970s, at the same time as the first oil crisis in 1979, a sudden growth of interest in photovoltaic (PV) electricity in the United States, Japan, and Europe was at the origin of a sudden development of research activities in the whole field of PV materials, devices, and processes.

Silicon was considered the material of choice for PV applications, by virtue of its excellent electronic properties, the appropriate match of its energy gap to solar radiation, and its environmental stability. However, since its cost was the major ingredient of the solar cell cost and then of the PV modules, R&D activities in the semiconductor materials field were typically addressed to the investigation of processes capable of reducing its cost while getting solar-grade quality, as discussed and defined in Chapter 1.

After few years of booming activity, at the beginning of the 1990s research on solar silicon was virtually abandoned, but 20 years later, together with the increase in global demand for silicon associated with the explosive growth of the PV market, R&D activities in solar silicon found a new impetus that is not yet exhausted.

The main aim was, and still is,* to develop a low-cost industrial process capable of competing in quality with the classical Siemens process, without passing through the chlorosilane route, with its high energy costs and potential environmental problems.

Most of the processes investigated so far are either variants or potential improvements of the classical metallurgical silicon (MG-Si) route and of the Siemens process, for which full industrial expertise exists, or processes aimed at MG-Si purification using chemical and physical methods.

This chapter deals with early attempts and recent progress with regard to the development of processes based on the direct MG route, using pure precursors, while the MG silicon refining processes will be discussed in Chapter 3, and gas-phase processes alternative to Siemens in Chapter 5.

Chapter 4 will be, instead, fully devoted to the Elkem process, the first and still unique industrial process based on the purification of standard MG-Si.

Particular emphasis in this chapter is devoted to the causes that determined the lack of success of early attempts and the challenges intrinsically associated with the direct MG-Si route, in view of the production of a low-cost solar silicon.

2.2 THE MG SILICON PROCESS

MG-Si is produced in millions of tons/year in multiMW submerged electrode furnaces, using crushed natural quartz as the silicon source and a blend comprising coke, coal, charcoal, and wood chips as the reductants [1–5].

It is a relatively low-energy-demanding process (12 kWh/kg), at least in comparison with the classical Siemens process based on the reduction of trichlorosilane (180 kWh/kg) and for this reason ideally competitive with it, provided it could lead to a solar-grade silicon (SOG-Si).

It is based on the carbothermic reduction of quartz with carbon, formally described by the following equation:

$$SiO_2 + 2C \rightarrow Si + 2CO \tag{2.1}$$

carried out on a vertical cylindrical furnace in which an heterogeneous mixture of quartz and coal lumps is thermally activated and reacted by means of a submerged arc, driven at the industrial scale by three graphitic electrodes. The upper wall of the furnace is made of ceramic bricks, while the bottom, where the liquid silicon is collected, consists of graphite or carbon-based materials.

Different from the carbothermic reduction of iron ores, the reduction of quartz occurs in the presence of stable (CO and SiC) or metastable (SiO) intermediate species, which participate to a number of binary equilibria

---

* The Silicor project described in Chapter 3 is one of these novel initiatives.

$$SiO_2 + Si \rightleftharpoons 2SiO \qquad (2.2)$$

$$2SiO_2 + SiC \rightleftharpoons 3SiO + CO \qquad (2.3)$$

$$SiO + 2C \rightleftharpoons SiC + CO \qquad (2.4)$$

$$SiC + SiO \rightleftharpoons 2Si + CO \qquad (2.5)$$

$$SiO_2 + C \rightleftharpoons SiO + CO \qquad (2.6)$$

$$SiO_2 + CO \rightleftharpoons SiO + CO_2 \qquad (2.7)$$

In addition to these binary equilibria, an irreversible condensed-phase reaction could also occur, leading to the direct formation of SiC

$$SiO_2 + 3C \rightarrow SiC + 2CO \qquad (2.8)$$

Reactions (2.7) and (2.8) could play a critical role, however, only in the case of granular charges of silica and carbon [6], as will be seen in Section 2.3.

The process may be better understood by looking to the equilibrium diagram reported in Figure 2.1, where two invariance points fix the coexistence of three condensed phases [7].

Although the matter is still under discussion, there is a general agreement that the process goes first through the primary formation of SiC in the upper (colder) region of the reactor, at and above the temperature of the first invariance point at 1512°C (1785 K).

The formation of SiC does not occur as a condensed-phase reaction involving the direct interaction of carbon with quartz (see Equation 2.8), but it forms as a result of the stoichiometric combination of reactions (2.3) and (2.6), working at a constant p_{SiO} pressure, buffered by the simultaneous occurrence of reaction (2.4) in the range of temperatures where SiO is stable against decomposition.

It is interesting to note that also the Acheson process, used for the industrial production of SiC, occurs via the same reaction paths, as the direct, condensed-phase reaction would be kinetically hindered in the presence of large particle sizes used in industrial reactors [8]. It should be noted that reaction (2.4) plays, also, a key role for the full recovery of the unreacted SiO coming from the bottom reaction zone (see later), which otherwise would be lost in the furnace fumes.

Since these reactions involve the simultaneous presence of gaseous (or vapor-phase) and solid reactants, their kinetic behavior depends on the morphology, porosity, and reactivity of the heterogeneous charge filling the upper portion of the furnace.

The formation of liquid silicon occurs near the bottom of the reactor, at and above the equilibrium temperature of 1811°C (2084 K), corresponding to the second

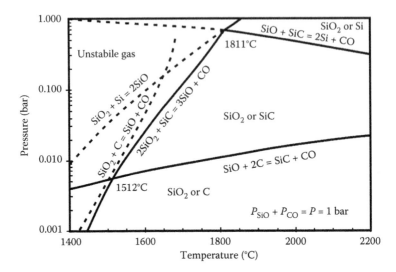

FIGURE 2.1 Phase diagram of the Si–C–O system: Reactions involving SiO and CO in the region of thermodynamic SiO instability (left corner of the figure) result in the condensation of SiO, if the system runs under equilibrium conditions. The pressure values represent the partial pressure of SiO. (After A. Schei, J. K. Tuset, and H. Tveit, *Production of High Silicon Alloys*, Tapir Forlag, 1998. Reproduced with permission of the author and of Tapir Forlag.)

invariance point, and involves the reaction of SiC, formed in the upper portion of the furnace, with unreacted quartz, according to the following equation:

$$SiO_2 + 2SiC \rightleftharpoons 3Si(l) + 2CO \qquad (2.9)$$

that corresponds to the stoichiometric combination of reactions (2.2) and (2.3), thus involving again the active role of SiO as a reaction intermediate.

In industrial furnaces, the silicon formation rate is governed by reactions (2.2) and (2.3) but is influenced as well by the recovery of unreacted, volatile silicon monoxide SiO which leaves the bottom reaction zone, where its partial pressure is very high, through the reactor-bed porosities. Recovery is associated either with its disproportionation to SiO_2 (reverse of Equation 2.2) or with its reaction with unreacted carbon at temperatures higher than the first invariance point, on the upper section of the furnace.

The MG-Si furnace is, therefore, a very complex semicontinuous reactor working in a temperature gradient, which needs to be optimized by a careful compromise of energy input and thermal conductivity and porosity of the charge. Woodchips add a specific contribution in the cooling of the very top of the bed, thanks to the evaporation of water that might be sprayed on them.

Owing to the impurities present in the raw materials and the high temperatures of the processes occurring in such a reactor, which favor strong interactions of silicon with the furnace materials and with the electrodes, MG-Si is a rather impure material.

TABLE 2.1
Average Impurity Concentration in Commercial MG Silicon

Impurity	Concentration (ppmw)
Fe	2000–3000
Al	1500–4000
Ca	500–600
B	40–80
P	20–50
C	600
O	3000
Ti	160–200
Cr	50–200

Its overall impurity content ranges, in fact, around 1%–2% [9] (see Table 2.1) well above any, even drastically relaxed, solar-grade quality.

The concentration of B and P dopants, in particular, is orders of magnitude higher than requested for solar applications,[*] and their removal could not be carried out using simple crystallization processes, as their segregation coefficients are close to 1 (see Chapters 1 and 3).

In addition, the reaction of SiO_2 with SiC (see Equation 2.9) leading to the liquid silicon formation, does occur in a carbon-rich environment, since the reactor walls in the bottom reactor zone are made of graphite or carbon, as are the electrodes. For this reason the reaction product is a liquid Si-C alloy, with a C content of the order of 500 ppmw. The carbon present in solutions rapidly reaches its equilibrium solubility value during smelting, when spontaneous cooling causes the segregation of SiC.

MG-Si is, therefore, well suited for metallurgical applications[†] or as the feedstock of the Siemens process,[‡] not for the direct growth of silicon ingots of PV use. In fact, for an impurity content larger than about 1000 ppmw, even after the preliminary removal of SiC, the planar growth turns out, short after the start, to a cellular structure, where all impurities are trapped as electrically active species.

Furthermore, it will result in a heavily compensated material, due to the presence of both donor and acceptor dopants, with severe consequences concerning its PV quality, as seen in Chapter 1.

The processes critically reviewed in this chapter have all in common the hypothesis that an advanced MG process should be able to produce a silicon feedstock of purity convenient, at least, for a first physical purification process, and therefore already low in its B and P content.

---

[*] They should stay, possibly, below 0.2 ppmw (see Chapter 1).

[†] As a component of light alloys or as a deoxidant in the iron metallurgy.

[‡] The Siemens process itself is a chemical/pyrometallurgical process which converts MG-Si in a mixture of silicon- and metal-halides. The latter are removed by distillation, while the silicon halides are reduced to polycrystalline silicon in hydrogen atmosphere.

This purification process should be carried out in a Czochralski (CZ) or a directional solidification (DS) furnace, where impurities are segregated in the liquid phase, leaving a single crystalline or polycrystalline ingot. This ingot should be of purity suitable to be used either as the material for a second, definitive growth of a single-crystalline or multicrystalline ingot, or directly usable, after wafering, for solar cells manufacturing.

It is, indeed, supposed that the use of a charge of very pure quartz and carbon reductants, extremely low in B and P would fulfill this requirement, in the hypothesis that furnace components and working tools are properly selected and used to avoid, or at least to limit, impurity contamination in an extremely high-temperature environment.

The analysis of the results of a number of these projects, carried out at the pilot scale, is illustrative of some intriguing challenges of a clean MG-Si process, which drastically limit its margins of success toward a solar-grade quality, unless paying extreme care to all process steps. We will see, in detail, that the main challenges of these processes depend on the difficulty of a smooth operation of the furnace without the use of conventional carbonaceous reductants (coal, charcoal, and wood chips) and in the containment of the spontaneous* contamination effects at the high temperatures of the MG furnace operation.

2.3   THE DOW CORNING–EXXON–ELKEM PROJECT

The Dow Corning Corporation, at the Solid-State Research and Development laboratory in Hemlock, started a research activity in the year 1976 addressed at the production of SOG-Si from purified MG silicon† [10–13] jointly with Elkem (Norway) and Exxon.

As most of the challenges associated with the production of SOG-Si with the MG route were originally observed (and only partially solved) in the course of this project, a detailed account of the project's results will facilitate the analysis of other research activities carried out in the same period by other companies.

The hint toward this attempt was the consideration that a substantial reduction of solar cells cost ($8.5/Watt in 1980) could be obtained with a reduction of the silicon wafers cost, since it was the major ingredient of the module cost at that time.

Furthermore, the rationale behind it was also the hypothesis that the use of commercially available, pure raw materials (natural quartz and carbonaceous reductants as such or properly purified) in a conventional MG silicon furnace, carried out under the best environmental conditions, would result in a material that could be used as PV feedstock after one or two crystallization processes.

The following sequence of process steps was foreseen:

1. Selection and/or purification of raw materials (quartz and carbonaceous reductants)

---

* Thermodynamics plays a major role, since all interaction reactions are favored by high temperature.
† L.P. Hunt was a good friend of S.P. Soon after the end of the Dow project he left the company to join Exxon-Mobil and Kerr-McGee, from which he retired in 1994. Then he received a B.A. in 1999 in Pastoral Ministry and was ordained a permanent deacon in the Roman Catholic Archdiocese of Oklahoma City where he served the parishes of St. Monica and St. John the Baptist Catholic Church in Edmond until March 2014, when he died.

2. Carbothermic reduction of quartz in a direct arc reactor (DAR) and production of DAR-Si
3. Pyrometallurgical purification of the DAR-Si obtained in step 2 (at a price goal of $14/kg (1980 dollars)
4. Physical purification using a CZ furnace* and growth of a polycrystalline ingot[†]
5. Growth of singe crystal (CZ) ingots using the material produced in step 4
6. Wafering
7. Solar cells processing

In the absence of an accepted solar grade (SOG) definition, on which worldwide research was carried out in the same years (see Chapter 1), the purity requirements of DAR silicon were estimated by assuming that a single crystallization process should reconduct the DAR silicon to a SOG quality.

The latter was defined on the base of an internal solar silicon definition (see Table 2.2), coming from the known experimental effects of dopant impurities and metallic impurities on cell efficiency, which fitted, however, fairly well with the results of Davis et al. [14] and of Pizzini et al. [15–17] (see Chapter 1).

For metallic impurities, the maximum allowable concentration in DAR silicon was consequently calculated (see Table 2.2 again) on the base of the their assumed SOG concentration and on their effective segregation coefficients $k_{CZ} = (c_i^s/c_i^l)$, for a crystallization process carried out using a CZ furnace. The selected segregation coefficients were a mix of literature values that fit relatively well with today's accepted segregation coefficients [18].

Different arguments were applied to B and P, since for these impurities a segregation purification process would be ineffective, due to their segregation coefficients being close to 1. Boron requirements were inferred from the literature data showing that solar cell efficiency would peak between 0.1 ohm-cm and 0.3 ohm-cm, i.e., for B concentrations between 9.6 and 1.7 ppma, respectively. A maximum concentration of 1.7 ppma (0.65 ppmw) for boron was chosen as the target. Consequently, the phosphorous concentration was set at 0.5 ppma (0.55 ppmw) in order to avoid strong boron compensation. It should be noted that the SOG concentration of B looks sensibly relaxed with respect to today's criteria, as light-induced degradation (LID) effects were unknown at the time of the Dow Project.

It could be observed in Table 2.2 that the calculated values of some critical impurities (Al, B, Fe, and P) already stay much lower than that in MG-Si. However, the maximum allowable concentration of Cr, Cu, Fe, Ni, and Zr reported in Table 2.2 was considered above any realistic level, since in these conditions the regular growth would turn out to a cellular growth.

---

* Also directional solidification (DS) was attempted, but due to the high concentration of impurities in DAR silicon the DS purification failed totally.

[†] Previous experience showed that CZ growth from a melt of MG silicon results in a polycrystalline ingot.

TABLE 2.2

Assumed Allowable Impurity Content, Segregation Coefficients (k_{CZ}) Using CZ Growth and Calculated Allowable Impurity Content in DAR Silicon, Compared with the Impurity Content of an Industrial MG-Si Sample

Impurities	Impurity Content in MG-Si (ppmw)	Assumed SOG Concentration (ppma)	Segregation Coefficient k_{CZ}	Calculated DAR-Si Allowable Concentration (ppma)	Calculated DAR-Si Allowable Concentration (ppmw)
Al	4300	0.2	2×10^{-3}	100	96
B	36	1.7	0.8	2	0.8
Cr	310	4×10^{-3}	1.1×10^{-5}	400	700
Cu	90	0.8	4×10^{-4}	1000	2000
Fe	8000	6×10^{-3}	6.4×10^{-6}	900	2000
Mn	200	2×10^{-3}	1.3×10^{-5}	200	300
Ni	80	1	3.2×10^{-5}	30,000	70,000
P	44	0.5	0.35	1.4	1.6
Ti	410	6×10^{-5}	2×10^{-6}	30	50
V	160	8×10^{-5}	4×10^{-6}	20	40
Zr	30	2×10^{-4}	$<1.5 \times 10^{-7}$	>1000	>4000

Source: From Dow Corning Final Report.

For this reason the maximum impurity concentration of these impurities was empirically scaled down and reported in the first column of Table 2.3, where also the data concerning Ca, which is a constantly present impurity in MG-Si, are considered.

From the first column of Table 2.3 one is able to forecast the allowable impurity concentration in the raw materials, assuming that the stoichiometry of the main process occurring in an arc furnace is given by quartz reduction reaction (2.1) and that metallic impurities are present as oxides in the silica matrix.

The calculated stoichiometric values should be, however, corrected for an impurity transfer coefficient η that accounts for the effective yield of impurity recovery in the course of the reduction reaction occurring in the arc furnace. The η values reported in Table 2.3 were independently measured on MG-Si samples produced in a 30 MW industrial furnace and are shown (as % recovery values) in Figure 2.2 to be proportional to the vapor pressure of the elements.

The data of Table 2.3 were also used to carry out the selection of quartz and carbonaceous reductants.

At least two available sources of commercial quartz were found of satisfactory quality (see Table 2.4).

Charcoal, the most convenient commercially available reductant,[*] and carbon black (CB) were selected as possible reductants, since coke, lignite, and woodchips

---

[*] Based on industrial practice of arc furnaces management.

TABLE 2.3

Calculated Maximum Allowable Impurity Concentration in Quartz and Carbon to Fit with the DAR-Si Purity

Impurity	DAR-Si Calculated (ppmw)	Impurity Transfer Factor (η)	SiO$_2$ (Calculated) ppmw	C (Calculated) ppmw
Al	100	0.8	60	50
B	0.8	0.66	1	0.5
Ca	600	0.5	600	500
Cr	700[a]	0.67	500	500
Cu	900[a]	0.73	600	500
Fe	800[a]	0.8	500	500
Mn	300	0.6	200	300
Mo	1400[a]	1	700	1000
Ni	800[a]	0.82	500	500
P	1.6	0.14	5	5
Ti	50	0.82	30	35
V	40	0.95	20	25
Zr	1300[a]	1.0	600	750

[a] Scaled down concentrations.

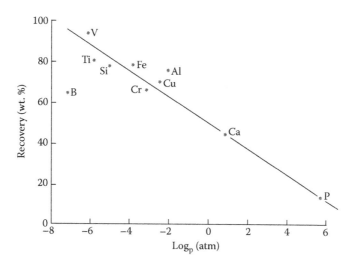

FIGURE 2.2 Dependence of the transfer coefficients η on the vapor pressure of the elements at the melting point of silicon. (After R. Davis et al., 1982. Effect of impurities and processing on solar silicon solar cells, Final Report, JPL no. 9850, Contract 954331, February 1982, pp. 1–226. Courtesy JPL.)

TABLE 2.4
Impurity Content of Two Commercial Sources of Quartz

Average Impurity Content (ppmw)	Type I (Clear)	Type I (Milky)	Type II	Target
Al	20	20	20	60
B	<1	1	2	1
Fe	<10	20	5	500
P	<10	<5	2	5

TABLE 2.5
Impurity Content in Commercial Charcoal before and after Fluorocarbon Treatments at Different Temperatures

Impurity (ppmw)	As Available	Purification 1550 (°C)	Purification 1700 (°C)	Purification 2000 (°C)	Purification 2500 (°C)	Target
B	21	27	8	15	5	5
Ca	8800	30	90	30	30	500
Mg	200	<5	<5	<5	<5	
Mn	120	<10	<10	<10	<10	300
P	230	25	160	20	<10	5

were discarded for their exceedingly high impurity content. Also the P content of charcoal (see Table 2.5) is particularly high, but it could reduce to values close to the target after an high-temperature purification treatment in a fluorocarbons atmosphere,[*] with satisfactory results common to all other impurities at temperatures above 2000°C. We will see later in this section that due to such a high-temperature treatment the charcoal reactivity decreases, with severe consequences on the furnace operation and silicon yield.

The impurity content of commercial CBs fully fitted, instead, with the quality targets and were therefore considered suitable as reductants, in the form of pellets or briquettes,[†] using sucrose or starch as a binder,[‡] which also increases the reactivity of CB.

CB powder could not be used, as such, in DAR experiments and must be pelletized to increase its micrometer-size particles to 5–25 mm-size pellets.

DAR experiments were carried out in single electrode furnaces of 50 KVA (DAR-50) (see Figure 2.3) and 200 KVA (DAR-200).

As the ceramic lining of the interior of the furnace was considered a powerful source of impurities, the electrode diameter was suitably chosen in separate

---

[*] A treatment that is currently used to purify graphite.
[†] Pellets and briquettes are used as synonyms in this chapter.
[‡] Both starch and sugar are extremely pure and very low in B and P.

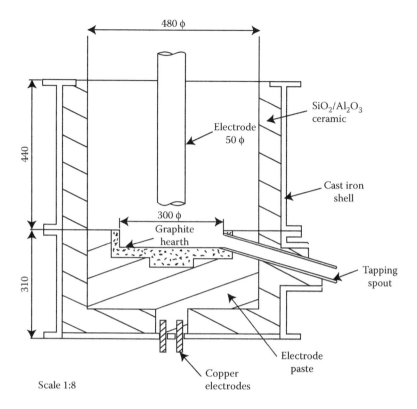

FIGURE 2.3 Construction drawing of the DAR 50 furnace (dimensions in mm). (After R. Davis et al., 1982. Effect of impurities and processing on solar silicon solar cells, Final Report, JPL no. 9850, Contract 954331, February 1982, pp. 1–226. Courtesy JPL.)

experiments in order to bring the high-temperature reaction cavity around the electrode (see Figure 2.4) sufficiently far from the furnace walls. In the conditions of Figure 2.4 the unreacted material around the cavity behaves, in fact, as a thermal shield that protects the ceramic bricks from high-temperature corrosion.[*]

DAR tests were carried out discontinuously, with a frequent dismantling of the furnace to investigate the nature of materials found in correspondence of the reaction cavity.

The productivity of both furnaces was low, with a maximum yield/test of 40 kg of silicon with the DAR-50 furnace and of 110 kg with the DAR-200 furnace, operated with charcoal as reductant. Reaction rates were low and the energy consumption very high, especially in the case of the DAR-50 furnace (from 40 to 160 kWh/kg). Better results were obtained with the DAR-200 furnace (>20 kWh/kg), but still unsatisfactory, also because the electrode consumption was exceedingly too high, up to more than 600 g/h.

---

[*] This design concept remains of interest, even today, as the integrity of furnace linings is limited by the conditions of furnace operations.

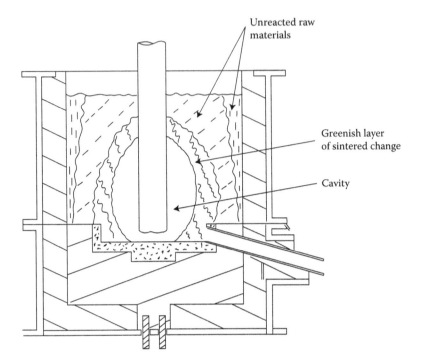

FIGURE 2.4 Cross-sectional view of the reaction cavity on the bottom of the furnace. (After R. Davis et al., 1982. Effect of impurities and processing on solar silicon solar cells, Final Report, JPL no. 9850, Contract 954331, February 1982, pp. 1–226. Courtesy JPL.)

The impurity content of the DAR-Si produced with purified charcoal as a reductant is reported in Table 2.6, which shows that Al, Ca,Fe, Mn, and Ti are systematically above the target.

Independently of the unsatisfactory Si-quality, the conclusion was that due to its low reactivity, activated charcoal could not be used for the commercial production of SOG-Si.

The results obtained using quartz and CB pelletized with starch and sucrose in DAR-200 experiments are reported in Table 2.7.

One can see that the use of sucrose as a binder leads to better results in terms of metallic impurity content as compared with those obtained with activated charcoal. Also the reaction rates (see Table 2.8) and the silicon production per run (128 kg) are the highest measured in the whole set of carbothermic tests.

B and P show massive deviations from the target in the majority of the runs, bringing out of SOG the whole silicon produced.

Another problem common to these experiments is that the energy consumption is a factor of 2 or 3 higher than that of MG-Si, see Table 2.8 again, and that in most cases the test should be stopped due to SiC formation on the bottom of the furnace, with the occlusion of the tapping spout, a clear sign that the carbothermic process does not occur under standard conditions.

TABLE 2.6
Average Impurity Content in DAR-Si Produced Using Activated Charcoal

Impurity (ppmw)	DAR 200	DAR 50	Target
Al	140–2800	40–300	100
Ca	500–7900	50–6500	600
Cr	<10	<10	700
Cu	10–45	5–300	900
Fe	1100–15,000	120–4000	800
Mn	40–650	<10	300
Ni	<10	<10	800
Ti	40–370	<10–60	50
V	10–30	<10	40
Zr	<10	<10	1300
B[a] ppma		12–14	1.7
P[a] ppma		9.7–16	0.5

[a] Measured in an ingot after CZ growth.

TABLE 2.7
Impurity Content in DAR-200 Experiments Using Carbon Black Pelletized with Starch (12, 13, and 14) or Sucrose

Impurity (ppmw)	DAR 200(12)	DAR 200(13)	DAR 200(14)	DAR 200 (16)	DAR 200(17)	After CZ Growth[b] (ppma)
Al	320	260	80	100	70	<1
Ca	80	360	40	40	50	(Mg) <0.003
Cr	<10	<10	<10	<10	<10	<0.003
Cu	<5	<5	<5	<5	<5	<0.015
Fe	2700	470	50	50	80	<0.03
Mn	<10	10	<10	<10	<10	<0.003
Mo	<10	<10	<10	<10	<10	<0.010
Ni	50	<10	<10	<10	<10	<0.03
Ti	<10	40	20	<10	<10	<0.005
V	<10	<10	<10	<10	<10	<0.003
Zr	<10	<10	<10	<10	<10	<0.012
B[a] ppma	12	11	13	9.6	11	11
P[a] ppma	4.5	5.4	4.5	5.6	0.45	<1

[a] Measured in an ingot after CZ growth.
[b] Originated from a DAR 200 material (17).

TABLE 2.8

Average Parameters for DAR 200 Experiments Using Various Reductants

Reductant	Time (h)	Silicon wt. (kg)	Average Reaction Rate (kg/h)	Average Energy Consumption (kWh/kg)
AC	29	45	1.9	29
AC + sucrose cubes	60	100	1.8	33
CB/starch	30	37	1.7	40
CB/sucrose	54	128	2.7	26

AC = activated charcoal.

We will see that this is a common condition of all the processes carried out with unconventional reductants.

Samples of DAR silicon were submitted to purification tests using both CZ and DS growth. DS was ineffective, as constitutional supercooling and cellular growth inhibited regular growth from the very beginning.

CZ purification runs better, due to the effective mixing of the melt [19], but only polycrystalline ingots were obtained, even with the use of single-crystalline seeds. Constitutional supercooling onset, and cellular growth occurs after the 77% of the ingot was grown also in the case of the CZ growth, showing the technical and economic limits of the Dow process.

The best material obtained by CZ growth of DAR 200 (see last column of Table 2.7) presents a metal concentration content similar to that found in electronic-grade (EG) silicon and a resistivity of 0.19 ohm-cm and was used to fabricate solar cells using routine industrial processes. The efficiency ranged between 10% and 13%, the highest value being comparable to that of solar cells currently fabricated in those years with EG silicon.

It will be seen in Chapter 4 that the DAR project, abandoned in the United States, was the early starting point of Elkem activities in Norway, and therefore the prototype of the modern solar silicon of MG origin.

2.4 THE HELIOSIL/EUROSOLARE PROCESS

In the early 1980s, as a spin-off of the research activities on semiconductor materials at the Central Research Institute[*] of Montedison (Novara-Italy), Heliosil started a research project on the production of SOG-Si from upgraded MG silicon. As in the case of the Dow project, Heliosil shared its R&D activities with two major Italian MG-Si producers, to whom the success of the initiative was certainly, at least partially, due, and with Eurosolare, the major Italian PV company at that time.[†]

---

[*] Istituto di Ricerche Donegani.
[†] S.P. was the Heliosil CEO. He is indebted to Massimo Rustioni and Daniele Margadonna, who were associated with the project.

The main goal of the project was to demonstrate the technical and economic feasibility of a MG process addressed at the production of SOG-Si, using low-B and -P, but commercially available raw materials, with a three-electrode, 270 kVA furnace, of structure and power comparable with industrial MG furnaces. As an internal standard, SOG-Si was defined as a material having a total amount of impurities less than 100 ppmw, with B and P ≤ 1 ppmw, each.

The aim was also to develop purification processes capable of bringing MG-Si to SOG, which were identified in acid leaching and crystallization processes. To this last scope a dedicated Bridgman furnace was designed and built, based on the innovative use of quartz crucibles coated with silicon nitride [20]. This technology has been used, later, in a joint initiative with Crystallox, addressed at the industrial production of multicrystalline silicon. It will be seen in Chapter 4 that the Heliosil–Eurosolare project, that did not find industrial development in Italy, found instead a partial continuation within the Elkem project.

Preliminary activities were addressed at the selection of commercial quartz and reductants of appropriate B and P content (see Table 2.9). At least two quartz sources were found compatible with a low B target, which could drop to 0.2 ppmw. CB was selected as the reductant, but the attempt to use SiC as a reductant was also carried out, as a means to see whether SiC (which is one of the intermediates in the reaction scheme of the carbothermic process) could be used in blend with SiO_2.

CB was used in the form of pellets or extruded blocks, with commercial sugar as a binder.

TABLE 2.9
Impurity Content of Quartz and Sand Samples Taken into Consideration

Element (ppmw)	Quartz 1	Quartz 2	Quartz 3	Quartz 4	Quartz Sand	CB1	CB2
Al	66	1700	275	40.7	132	21.2	27
B	0.2	4.5	nd	0.3	0.3	0.04	0.015
Ca	35	10	–	9.1	95	2.9	70
Cr	1.4	1.5	nd	0.9	0.1	0.9	2.7
Cu	0.7	nd	nd	0.3	0.6	0.03	0.4
Fe	37.6	200	<20	17.1	31.5	9	43
Mg	1.4	13	15	1.9	6	3.2	20
Mn	2.3	nd	nd	0.5	1	0.2	1.3
Mo	nd	nd	–	0.6	nd		
Ni	–	nd	–	–	1.0	0.8	0.6
P	0.3	5.0	2.0	1.5	3.0	0.3	0.078
Ti	10	35	40	6	60	1.1	1.0
V	0.7	nd	nd	0.1	0.2	0.8	0.2
Zr	0.1	nd	nd	0.2	4.0	0.02	0.02

Note: The impurity content of samples 1, 4, sand, and of CBs has been carried out with mass spectrometry, in the other cases with atomic adsorption spectroscopy. nd = not detected.

2.4.1   Set Up of the Furnace

A three-electrode furnace ($\Phi = 1200$ mm, h = 700 mm) of 270 kVA, already used for the production of silicon alloys, was selected as the main tool of the project. A lining made of high-purity graphite was applied to the entire depth of the furnace, to limit the impurity contamination of silicon from dirty ceramic bricks, but it was observed in preliminary experiments that such a lining would induce a remarkable loss of heat from the bottom reaction zone, with poor or none reaction yields. Therefore, the graphite lining has been left only on the bottom zone of the furnace (up to about 250 mm from the top) and the top zone was lined with high-purity aluminosilicate bricks.[*]

On this furnace the following tests were carried out:

1. Use of quartz sand and CB pellets
2. Use of quartz lumps and CB pelletized or extruded
3. Use of a blend of quartz lumps, SiC, and CB pellets
4. A limited number of tests of short duration carried out without silicon smelting

Only the most significant results of experiments that lasted more than 2 years are reported here next.

2.4.2   Results of Tests with Quartz Sand and CB as Reductant, Pelletized, or Extruded

A single run (HP1) was carried out using as the charge a blend of quartz sand, CB, and sugar, with a molar ratio $SiO_2/C = 2$, extruded in the form of $\Phi = 30$ mm, h = 10 mm cylinders,[†] using silicone oil as lubricant. After extrusion, the material was essicated in air at 120°C.

The furnace operation with an extruded material was rather unsatisfactory, with very high temperatures on top of the furnace and formation of hard crusts, difficult to be crashed even with steel tools.

Smelting was also difficult, due to the low temperature of the bottom zone, and therefore of the liquid silicon, which tends to freeze in the tapping spout. Opening with steel tools was often necessary, with dramatic degradation of the silicon quality in terms of Fe content. Loss of SiO vapors from the bed was also observed.

At the end of the test, when the interior of the furnace was dismantled, visual inspection showed that a block of well-crystallized SiC filled the entire volume of the crucible and that a strong corrosion of the ceramic lining[‡] occurred, due to the high temperatures on the top of the furnace.

The results of the analysis of the silicon obtained at the beginning of the test [Al 1145, B 12.5, Ca 174, Fe 4050, P 5. (ppmw)] confirmed the occurrence of corrosion

---

[*] Or preformed silica bricks at the very end of the experiments.

[†] Silicone oil was used to facilitate the extrusion process.

[‡] Corrosion of the ceramic walls was observed.

events in correspondence to the aluminosilicate lining and the detrimental role of the inappropriate handling of the tapping procedures.

As the furnace performances are smoother using pellets (see test HP2), it is difficult to judge the reason for this behavior, although one possible explanation could be that the essicating process induces the encapsulation of the mixture of quartz sand and CB in a silica layer, which favors the occurrence of a solid-state reaction between CB and silica with the direct formation of SiC, following Equation 2.8, to give a mixture of SiC and unreacted silica.

One could expect the following consequences:

- The formation of SiC, which is a good thermal and electrical conductor, induces a partial shorting of the electrodes on top of the furnace, responsible for the high temperatures observed at the very top of the furnace, of the corrosion of the aluminosilicate bricks, and of the low temperatures observed in the bottom reaction zone.
- As the charge remains in the total defect of carbon, due to the formation of SiC, the reaction

$$SiO + 2C \rightleftharpoons SiC + CO \qquad (2.4)$$

could occur only at graphite electrodes in the reaction cavity and part of the SiO formed in the bottom reaction zone leaves unreacted in the furnace fumes. Due to the high temperatures on the top of the furnace bed, probably also the disproportionation reaction

$$2SiO \rightleftharpoons SiO_2 + Si \qquad (2.2)$$

does occur only marginally.
- Most of the SiC formed by direct solid-state reaction on top of the furnace fills the bottom reaction zone unreacted, due to its low reactivity or to the low temperature of the bottom reaction zone.

A longer run of 6 days (run HP2) that led to a statistically significant output of 640 kg of silicon was carried out using a charge of pellets of quartz sand, CB, and sugar. Different from subsequent runs, the pellets were used as such, without further heat treatments needed to improve their mechanical strength. It was noted that working with a stoichiometric SiO_2/C ratio the furnace could not reach stationary conditions, unless adding an excess of carbon under the form of CB pellets.[*] Once the stationary conditions were set up, smelting could be conducted without particular difficulties for the entire duration of the run, but the use of steel tools was often needed to open the tapping spout, leading to massive impurity contamination of the liquid charge, even when a cup of SiC was used.

---

[*] Quite obviously, pellets of such kind lose carbon in the form of powder that ignites on top of the furnace.

TABLE 2.10

Energy Consumption and Impurity Contamination of Silicon Material Produced in Run HP2

Day	Energy Consumption (kWh/kg)	Silicon Produced (kg)	Fe	Ca	Al	Ti	P	B
1	73	52	4740	480	2950	860	42	14
2	32 (after C additions)	140	5700	370	2580	790	46	16
3	25	180	3960	540	4500	710	60	16
4	45	108	4000	140	2180	400	62	18
5	60	84	3230	110	1620	300	65	17
6	17	76	1920	33	2350	230	32	18

As in the case of the HP1 run, the top of the furnace was very hot, with the formation of hard crusts, which limit the vertical movement of the electrodes as well the escape of the reaction gases. These crusts were difficult to crash, unless steel tools with a silicon carbide cap were used.

Details concerning the silicon production yield and its impurity content (ppmw) in run HP2 are reported in Table 2.10.

One can see that the energy consumption is very high, as compared to that of conventional MG silicon, but compares with Dow values, and that the Al and Ca content also indicates in this case the onset of corrosion events of the aluminosilicate bricks, like in the HP1 run.

Fe, P, and B contamination arises, instead, from the mismanagement of the furnace with metallic and wooden tools, the latter responsible for the high P content.

The high temperature (T > 800°C) on top of the furnace makes the manual operation of the furnace difficult, and also causes an unfavorable thermal profile of the furnace itself.* It is probably caused, as in the case of the run HP1, by a partial shorting of the electrodes associated with the presence of a top layer consisting of electrically (and thermally) conductive SiC.

This situation is very different from that occurring in the upper zone of an industrial MG furnace, where the formation of SiC occurs only by reaction of carbon with SiO escaping from the hot zone, which is produced by reaction 2.2.

The SiC reaction formation, see Equation 2.4, is also the reaction which provides the main recovery of SiO vapors [4,21].

These results confirm the evidence empirically acquired with the Dow project that a carbothermic process carried out with sand and CB leads to furnace operations far from those occurring in a classical MG process. These results show also that a low-silicon yield is intrinsic to processes in which SiC forms in the upper zone of the furnace by a direct reaction of quartz with carbon.

---

* That tends to get the temperature of the liquid silicon close to the freezing point.

A last attempt carried out to optimize the furnace operation conditions was done in the run HP12, using pellets of quartz sand, CB, and sugar which were submitted to an heating stage in air at 900°C for 4 h* to consolidate the material and avoid carbon and quartz sand losses during the furnace replenishment operations.

The operation conditions of the HP12 run, which lasted 128 h, were smoother than in the HP2 run and did not present smelting problems. Also in this case, however, the top of the furnace bed was very hot, with visible loss of SiO vapors. The formation of crusts was observed as well, but the crusts were in general more brittle and easier to crash. The overall conversion yield (total mass of raw materials/mass of silicon) was fairly low (65%), associated with the escape of unreacted SiO in the furnace fumes,[†] with an average energy consumption of 27.0 kWh/kg.

The results relative to the impurity content along the entire run are displayed in Table 2.11, where the expected impurity content in the last column is calculated on the base of the total impurities content in raw materials, including the carbon electrode losses, for an experimental conversion yield of 65%.

It could be seen that the experimental B and P content is only marginally affected by the furnace operation, with a B content in the limits of a SOG but a too high P content. The corrosion of the aluminosilicate bricks leads to a modest influence on the Al and Ca content, and the iron and titanium contents remain close to their theoretical content. The deviation observed at the beginning of the test probably depends on the improper smelting management with unprotected steel tools.

The silicon quality is outside the SOG target, as the total impurity content is higher than 1000 ppmw, against a target at <100 ppmw.

TABLE 2.11

Impurity Content of Silicon Relative to a Series of Smelting Steps of the HP12 Run

Element (ppmw)	HP12/1	HP12/5	HP12/4	HP12/6	HP12/8	HP12/9	HP12/10	Expected
Al	620	380	575	365	500	370	715	490
B	2.87	2.68	2.06	2.00	3.5	2.29	4.01	2.84
Ca	200	90	155	115	125	90	220	130
Fe	1750	710	787	615	830	700	870	800
Mg	7.2	10	10	14	10	11	15	11.3
P	30	23	38	32	26	26	37	30
Ti	490	275	385	270	350	280	590	365

Note: Regime conditions were achieved starting from HP12/5.

---

[*] Apparently this temperature is exceedingly too high to induce stable cross-linking and air is detrimental.

[†] According to Andersen [4] the silicon yield increases if less SiO leaves the hot zone.

2.4.3   Results of Tests with Quartz and Silicon Carbide (Runs HP6 and HP10)

Few runs were also devoted to carbothermic reduction of quartz with SiC, in the same three-electrode furnace used in previous runs.

The potential advantage of the use of SiC, a reaction intermediate in the conventional MG process (see Section 2.3.1), as the reductant of SiO_2 instead of carbon, is the lower amount of CO produced per mole of silicon

$$SiO_2 + 2SiC \rightleftharpoons 3Si + 2CO \tag{2.10}$$

just 1/3 of that evolved in the direct arc reaction

$$SiO_2 + 2C \rightarrow Si + 2CO \tag{2.1}$$

and an expected lower energy demand per mole of silicon produced, even accounting for the energy used for the industrial production of SiC (about 11 kWh/kg), which is a very efficient process.

The SiC used was either a commercial silicon carbide or ad hoc synthetized in an Acheson furnace, using pure quartz sand and CB briquettes with sucrose as the binder. The silicon carbide blocks thus produced were crushed to a size ranging between 1 and 20 mm, and used to feed the furnace together with quartz nuts of a size ranging between 20 and 60 mm, which were obtained by grinding and HCl rinsing natural quartz blocks.

The run HP6 lasted 88 h was carried out with commercial SiC and led to a total amount of 600 kg of silicon, with an average energy consumption of 17.3 kWh/kg and average reaction yield of 75%. The run HP10 lasted 38 h, with an average consumption of 11.9 kWh/kg and a reaction yield of 95%. For both runs, energy consumption and production yield were much better than using quartz sand and CB pellets.

The furnace operation is, however, critical. This is mostly due to the high temperatures on the surface of the bed (see Figure 2.5) to the formation of hard crusts on top (which limit the vertical electrode movement as well the escape of the reaction gases) and the accumulation of slags (fused quartz and SiC) on the bottom of the furnace that make the smelting process often extremely difficult.

To avoid the contamination of the material during the furnace operation (smelting, breaking the crusts which form on the top of the bed) graphite tools or steel tools with a SiC cap were used along the entire duration of the run.

The impurity content of the silicon produced with the use of SiC as the reductant is reported in Table 2.12 and compared with the best measured in the run HP12, where the precursors were silica sand and CB. It appears that the overall quality of the silicon produced using SiC as the reductant is worse than that obtained using quartz sand and CB, also because of the worser quality of the precursors.

FIGURE 2.5 Picture of the furnace in operation with a charge of SiC and quartz lumps.

2.4.4 PURIFICATION EXPERIMENTS USING ACID LEACHING AND DIRECTIONAL SOLIDIFICATION

Purification experiments were carried out only on silicon produced using SiC and quartz (run HP6 and HP 10), by submitting the MG-Si first to an acid leaching process, after grinding to a size around 0.5–2 mm,[*] and then to a single or double crys-

---

TABLE 2.12

Impurity Content of Si in the HP10 Run, Compared with the Best Results of the Run HP 12

Impurity (ppmw)	SiC	Quartz	Run HP10	HP12/6
Al	156	31	340	365
B	2.1	<0.5	5.8	2.00
Ca	194	60	153	115
Fe	290	10	986	615
Mg	45	15	27	14
P	5.6	5.2	6.1	32
Ti	96	5.2	382	270

Note: Also the impurity content of the precursors SiC and quartz is presented. The impurity content has been measured with ICP-ES measurements as in the previous tests.

---

[*] Silicon powder presents moderate pyrophoricity when the grain size is below 10 μm, limiting the purification efficiency of acid leaching. The solution of this problem has been found by alloying silicon with calcium [22,23], which will be discussed in Chapters 3 and 4.

FIGURE 2.6 Section of the ingot W263.

tallization process in a DS furnace, originally designed for this scope.[*] Leaching was carried out to remove the major amount of metallic impurities, thus to avoid growth degeneration associated with supercooling and cellular growth in the crystallization process, that actually systematically occurs in the absence of leaching.

The ingot (W263) (see Figure 2.6) obtained from the material coming from the run HP6 (SiC and SiO_2 as precursors) was multicrystalline, with large grains vertically oriented, without evidence of cellular growth. The impurity concentration is reported in Table 2.13, as a function of the relative height in the ingot.

One could see that, as expected, Al, Fe, and Ti concentrate on top of the ingot while the B concentration remains almost constant. Ti is very efficiently removed for the bulk of the ingot, while Al remains well above 1 ppmw in the whole set of samples.

TABLE 2.13

Impurity Concentration Profile in the Ingot of First Solidification of MG Silicon from Run HP6 (SiC and Quartz Lumps)

Impurity (ppmw)	Starting Material	26	30	50	52	54	92	96
Al	360	3.0	12.3	1.4	2.5	11.7	7.0	198
B	5.4	4.0	4.5	5.1	4.8	5.1	5.0	6.7
Ca	71.3	18.1	124	5.8	55	12.7	2.3	6.1
Fe	1230	7.7	11.3	<1	<1	1.3	<1	652
Mg	12.1	12.7	18.3	<1	<1	7.4	<1	<2
P	16.2	2.0	3.2	<1	2.0	1.8	1.8	7.3
Ti	140	<1	<1	<1	<1	<1	1.5	80.4

---

[*] In the frame of the development of the DS furnace, also the original use of quartz crucibles lined with silicon nitride was applied.

TABLE 2.14

Impurity Distribution in the Ingot W 271

Element (ppmw)	Starting Material[a]	9	11	15	50	56[b]
Al	168	6.6	8.8	7.5	4.9	275
B	2.7	2.8	2.9	2.8	2.6	3.1
Ca	81.5	126	84	40	183	330
Fe	425	3.7	5.6	4.8	18	640
Mg	31	34	28	15	12	31
P	4.5	3.8	6.0	5.2	5.3	16
Ti	160	<1	<1	1.5	<1	240

[a] MG silicon after leaching.

[b] Top of the ingot.

The behavior of Ca and Mg is anomalous, with a variable concentration along the entire height of the ingot.

Also the material obtained in the run HP10 has been submitted to a purification test, consisting in a standard leaching and a DS growth (ingot W271).

The impurity distribution on samples cut in a vertical section of the ingot W 271 is reported in Table 2.14. It compares with that of the previous example, and shows, in fact, that Mg and Ca could not be removed efficiently in spite of a segregation coefficient of 2.3×10^{-3} for Mg and 1.6×10^{-3} for Ca* [24]. Both impurities sit in substitutional positions of the silicon lattice and their solubility, respectively, is 1.5×10^{19} cm^{-3} and 6×10^{18} cm^{-3} for Mg and Ca.

One of the reasons of their incomplete removal could be that both present an important retrograde solubility behavior, leading to supersaturation conditions and to retrograde melting [25]. The main point, however, is that Ca and Mg in silicon behave as effective getters for oxygen, in view of their values of free energy of formation of the oxides that are in excess of that of SiO_2 (see Ellingham diagram of Figure 2.7). They should be, therefore, present as Me-O complexes or as MeO precipitates and could not be efficiently removed during the crystal growth.

Even after a second crystallization the Al, Ca, and Mg concentration remain above 1 ppmw, leading to a best value of the diffusion length of 50 μm. Consequently, the best solar cells manufactured with silicon obtained from the carbide route gave an average efficiency of 6.5%, to be compared with an efficiency of 10% measured on cells manufactured with off-grade polycrystalline silicon [27].

2.4.5 CONCLUDING REMARKS

The results of the experiments carried out within the Heliosil/Eurosolare project show that the furnace operation with CB–quartz sand pellets is a major challenge.

---

* When measured with a CZ growth.

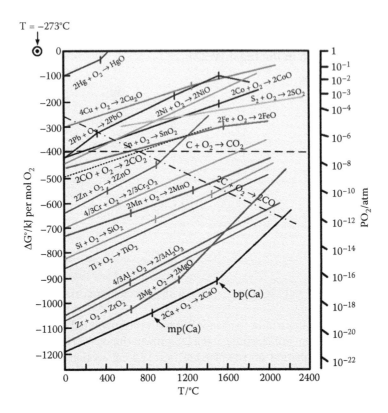

FIGURE 2.7 Ellingham diagram for common oxides. (After Y. Shen, 2015. Carbothermal synthesis of metal-functionalized nanostructures for energy an environmental applications. *J. Mater. Chem. A*, 3, 13114–13118. Reproduced by permission of The Royal Society of Chemistry, Licence nr 3720150705886, Oct 1, 2015.)

High surface temperature and crusts formation due the direct synthesis of SiC on top of the furnace and SiO escape are the main concerns that appear to be intrinsic to the use of CB–SiO_2 pellets or SiC as the reductant. As will be seen in Chapter 4 with the Elkem process, CB pelletized with sucrose is instead an excellent reductant when used in the presence of coarsely granulated quartz.

Independent of the use of CB or SiC as reductants, the energy consumption is a factor 2 higher than that of an industrial MG furnace and the reduction yield too low.

B content is not a problem, as it only depends on its content in the raw materials. P, Al, Ca, and Mg contamination was, instead, a problem that was not brought to solution.

Apparently, there is not a great advantage in the use of very pure precursors in terms of metallic impurities,[*] as the furnace operation brings the quality of the produced material not too far from that of the commercial MG-Si.

---

[*] Boron and phosphorous content remains instead essential.

2.5 THE SIEMENS ADVANCED CARBOTHERMIC PROCESS

Among the number of companies that were involved in the production of solar silicon by a direct route, Siemens AG[*] also attributed to the MG route a great potential impact for important cost reduction, provided the required purity could be obtained [28,29]. Different from previous projects, for which research reports were freely available, only open literature papers are available, where detailed process data are almost totally absent [30–34].

For this reason it is difficult to draw a conclusion about the technical feasibility of the process, also because the results are severely conflicting with those previously reported, where the challenges are considered superior to the benefits.

As for the previous projects, the activities of the ACR process were addressed at

1. The selection/preparation of extremely pure precursors
2. The operation of the furnace under high-purity conditions
3. The quantitative removal of SiC from silicon
4. The final silicon processing

Since the solar quality of the silicon produced with a carbothermic process critically depends on its B and P content, in the hypothesis that metallic impurities could be removed by segregation in the final process of ingot growth, the precursors (quartz and carbon) should be as low as possible in B and P content.

A B- and P-concentration of quartz below 0.2 ppmw was chosen as the goal, definitely lower than the B and P content in the quartz used in the Dow Corning project and in most commercial quartz.

Considering, furthermore, that high-grade natural quartz are expensive and that their B and P content is eminently variable and dependent on the quartz ore, the solution adopted was the chemical purification of an inexpensive commercial quartz,[†] done by reacting it first with glass-forming oxides (Na_2O, CaO, and Al_2O_3) to produce a fused glass. Glass fibers spinned from the glass melt are eventually leached with hot HCl in order to dissolve all the impurities present as oxides, leaving only amorphous silica, that is insoluble in HCl. The impurity content of this material fits the goal, as the B- and P-concentration is less than 0.2 ppmw, Mg < 0.1, Ca < 1, Al < 20 ppmw while the transition metal impurities are less than 1 pmw.

Pellets of CB leached with HCl were, instead, used as high-purity reductant.

These materials were used as the charge of a 70 kW furnace lined with high-purity graphite.[‡] Also the electrodes were manufactured with high-purity graphite. The high-purity silicon thus produced was p-type, with a B concentration of 5×10^{16} cm^{-3} (1 ppma) and a P concentration <10^{16} cm^{-3} [35].

---

[*] Already proprietary of the process for the production of electronic-grade silicon with a gas phase process.

[†] Chemical purification of quartz to be used as precursor of solar silicon is a process adopted also in the NEDO program for the production of *water glass silica* [36] and adopted and patented by RSI [37]. (S. Amendola, Method for making silicon for solar cells and other applications WO/200/106860).

[‡] We have already shown that a furnace entirely lined with graphite does not work properly, as too much heat is transferred by the thermally conducting walls.

The results of this experiment carried out with the small furnace were confirmed using a three-electrode furnace of 500 KVA, lined with pure graphite.

Further processing of silicon was carried out by a dual step process, of which the first was devoted to carbon removal and the second to ingot growth using a CZ furnace.

Carbon removal was carried out by remelting the material in a graphite crucible, holding the temperature slightly above the melting temperature of silicon. In these conditions carbon excess segregates in the form of small particles of SiC, which are allowed to grow until they sink to the bottom of the crucible due to their larger density as compared to that of silicon. Alternatively, SiC particles were removed by filtering.[*]

After carbon removal the material could be converted to a single crystal ingot by a single CZ pull.

Wafers cut from the second-generation ingots[†] (dislocation free, $\emptyset = 100$ and 125 mm, $R = 0.5$–0.4 Ω cm, $L_D = 50$ μm) were used as substrates for solar cells manufactured in a commercial process.

The best efficiency values obtained under Air Mass 1 (AM1) conditions were 12.2% for 100 mm cells, comparable to the efficiency of solar cells manufactured with EG silicon [38]. To the present authors' knowledge, the process did not reach industrial development.

2.6   THE SOLSILC PROCESS, OR A DIRECT ROUTE FOR THE PRODUCTION OF HIGH-PURITY SOG-Si

2.6.1   HISTORICAL GROUNDS

The Solsilc project was initiated in 1998 by Kværner and the Dutch company SunErgy AS. Kværner owned a process for production of pure carbon black (CB) and SunErgy was a building development company with interest in integrating PV in buildings. Knut Lynum from Kværner met the CEO and owner Bienno Wiersma of SunErgy in Cape Town where both companies were present as sponsors of Volvo Ocean sailing boats race. During the event, they called Aud Wærnes at Sintef who could confirm that CB probably could be used as a reductant for the manufacturing of SOG-Si. The decision to start a project was then made in 2000 and the EU/FP5 sponsored project Solsilc[‡] was then started. In addition to Sintef, Kværner, and SunErgy, ECN with competence in solar cells and Scanarc with competence in plasma technology joined the team.

The project was developed through a series of European projects. In 2005, the Norwegian silicon producer Fesil joined the team. Fesil was, in fact, positively impressed with the project and in 2007 the company Fesil–SunErgy was founded. The company was owned 51% by Fesil and 49% by SunErgy. A new pilot-scale factory was built in 2009 with pelletizing equipment, a 3 MW arc furnace, melt refining

---

[*] Both these processes are well known from the technical literature, but, difficult (not impossible) to be carried out properly at the pilot and industrial scale. Today suitable filters in porous SiC are commercially available.

[†] Probably arising from a second crystallization of primitive ingots.

[‡] ERK6-CT-1999-00005.

furnaces, and DS furnaces. The raw materials were pure CB and quartz powder delivered from The Quartz Corporation.

In 2010, Fesil sold its MG-Si plant at Holla to Wacker and concentrated only on Si for PV. The results of the research activity were good and technically consistent and the decision was made to develop a process at industrial scale that could produce PV-grade silicon. Shortly after this decision PV silicon prices dropped and an economical evaluation of the project costs brought the conclusion that the cost of high-purity quartz was too high. The process was then afforded with lower-grade quartz, and also in this case the problems arising due to the high Al content was solved with a melt refining process. In a further search for low-cost high-purity quartz Fesil came in contact with the German company Evonic that produced large amounts of high-purity silica as a by-product. Evonic bought Fesil–SunErgy in 2011 and after 1 year the project was closed down.[*]

2.6.2 THE DIRECT CARBOTHERMIC ROUTE: MAIN CONCEPTS

In the Solsilc process, SOG-Si is made by direct carbothermic reduction of high-purity, low B and P quartz.

As in the case of the other processes discussed in the previous sections, the purity of the reductants, consumable electrodes, and furnace lining[†] were also taken into proper account when setting up the criteria for designing the various steps of the process, without neglecting that the trade-off between purity and price is the key to final success.

In a direct carbothermic route, the liquid silicon leaves the reduction furnace at a temperature between 1600°C and 1800°C. At this temperature the equilibrium level of dissolved carbon is 500 ppmw, according to the C-Si phase diagram [39], which segregates as SiC upon cooling. In addition to the dissolved carbon, SiC, C particles, and oxide films can be found in the tapped material. For the final application of SoG-Si to solar cell fabrication, all particles above a certain size should be removed. The SiC particles are hard and may present electrical conductivity as a consequence of co-precipitation of metallic impurities. This can create problems in the wafer sawing process and induce local short circuits in the solar cells. As in addition to carbon, metallic impurities are also present, advanced refining techniques are necessary to clean silicon, where high process temperatures and high reactivity toward refractories are extra challenges.

2.6.2.1 The Raw Materials

2.6.2.1.1 Quartz and Reductants

In a conventional MG process, lumpy quartz, coal, and woodchips[‡] are used, but these materials are not clean enough. On the other side, pure quartz and pure C can only be found as fine powders in the sub mm range.

---

[*] The Solsilc activity was supported by the following EC-funded projects: 2001 Solsilc (process design and development); 2003 SPURT (up scaling for refining); 2004 SiSi (optimization, recycling of kerf); 2006 FOXY (refining of solar grade silicon).

[†] Due to the high process temperatures, interaction of reactants and liquid silicon with furnace lining is a direct cause of impurity contamination.

[‡] Added to control the bed porosity and to maintain cool the top of the furnace.

Since the carbothermic furnace (see Section 2.2) is a semi-continuous reactor with relatively large flow rates of CO, SiO, and CO_2 in counter-flow with the coarsely grained raw material and that woodchips cannot be added due to their high phosphorous content, the main challenge to make the furnace work is to have a good process for agglomeration of the raw materials.

Industrial standard processes for agglomeration (pelletizing and briquetting) were considered and tested. Clean binders were also identified and the pelletizing process was optimized with a mix of selected binders.

Only a few deposits of high-purity quality quartz are available in the world and this may be a problem for the future evolution of the solar silicon market. There are no strict rules to find new ones, but some indications exist on how high-purity quartz is formed and where it is likely to occur [40].

The currently mined high-purity quartz deposits are pegmatite (Drag, Norway; Spruce Pine, USA) or hydrothermal veins (Saranpaul, Russia). Undeveloped deposits of high-purity quartz in Norway are hydrothermal veins (Svanvik) and kyanite quartzites [41].

It is industrial practice to refine high-purity natural quartz down to very low impurity levels by crushing and acid cleaning. The remaining impurities in the quartz powder are in the ppmw or sub-ppmw range (see Table 2.15).

CB does not give, instead, a significant contribution to the impurity level of the raw materials.

2.6.2.1.2 Electrodes and Furnace Body

High-purity electrodes were used, especially developed for the Solsilc process.

A special furnace body lining* was developed in cooperation with Sintef to reduce the impurity levels and tested in the pilot-scale furnace. The lining has the following composition:

- 80%–95% (w/w) of a silicon carbide product containing 1.5 ppm (w/w) or less boron and 3.0 ppm (w/w) or less phosphorus

TABLE 2.15

Impurity Concentration of Selected High-Purity Quartz Products

	NC4A	NC4X	NC1CG	Iota 4	Iota 6	Iota 8	Quartz
Al	14.0	14.0	25.0	8.0	8.0	7.0	17.0
B				0.04	0.04	<0.04	<1.0
Cr	<0.03	<0.03	0.01	<0.05	<0.05	<0.003	<2.0
Cu	<0.03	<0.03	0.01	<0.05	<0.05	<0.002	<2.0
Fe	0.3	<0.06	0.8	0.30	0.15	<0.03	<5.0
P				0.05	0.05	0.05	<0.5
Ti	1.2	1.2	3.0	1.4	1.4	1.2	1.4
Supplier	Norwegian Crystallites			Unimin Corporation			Dr. Jacobs
References	[41]			[42–44]			[40]

* Patent EP 2 530 051 A1.

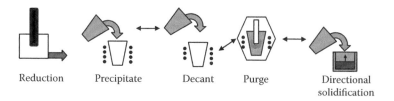

| Reduction | Precipitate | Decant | Purge | Directional solidification |

FIGURE 2.8 Schematic drawing of the Solsilc concept. (Reproduced from E. Øvrelid et al., 2006. *Silicon for the Chemical Industry VIII*, 223–239. With permission of Tapir Verlag.)

- 5%–20% (w/w) of a silicon product containing 1.5 ppm (w/w) or less boron and 3.0 ppm (w/w) or less phosphorus
- 0.1%–5% (w/w) of a binding agent, such as a resin, e.g. a synthetic resin

2.6.3 THE SOLSILC CONCEPT

After reduction of quartz with carbon in the electric arc furnace, the metal goes through several refining steps aimed at the removal of SiC particles, of excess P, and of the metallic impurities. This last process was carried out with a DS furnace. A principle sketch of the Solsilc concept is presented in Figure 2.8.

Precipitation by settling was used to reduce the carbon level from the material spilled from the furnace. Purging was used to remove C and Al and DS was mainly used to reduce the content of Al and Ti, which are the main contaminants (see next section). The impurity content in the Si ingot obtained in the solidification process determines the final quality of the product.

2.6.4 QUALITY OF SILICON OBTAINED IN THE SOLSILC PROCESS

Due to the low metal content in the raw material, the tapped silicon was already significantly pure, with Ti and Al as the main contaminants.

In order to forecast the behavior of such material and to establish purity targets, early in the project, multicrystalline ingots were grown starting from EG silicon with addition of Ti and Al to simulate the final product. The material was used as substrate for solar cells and the solar cell efficiency was thereafter measured as a function of the impurity levels, as shown in Figure 2.9.

Table 2.16 reports the calculated and experimental* impurity concentrations, the target values to be satisfied for a SOG feedstock, as compared with the impurity content of commercial low-grade electronic silicon and of Ekem-Silgrain silicon, on which will be discussed in Chapter 4.

The results reported in Table 2.16 show that a silicon ingot of reasonable quality, including the B and P concentration, could be obtained after one single DS growth. The average lifetime was shown, however, to be very low, associated also with a relevant density of oxygen precipitates and to a high iron content [47].

---

* The results from the Foxy project [49] are reported in this table.

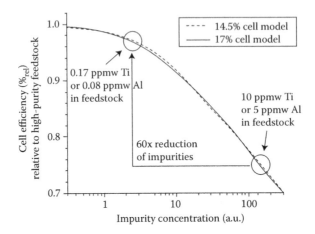

FIGURE 2.9   Effect of Ti and Al on solar cell performance. (From L. J. Geerligs et al., 2005. Specifications of solar-grade silicon: How common impurities affect the cell efficiency of MC-Si solar cells. *20th PVSEC*, Barcelona.)

TABLE 2.16

Impurity Levels in ppmw after Refining Silicon Produced by the Direct MG Solsilc Route and Target Values of Impurity Levels Required to Make Solar Cells

Impurities	Fe	Ti	Al	P	B	C
Solsilc final product estimated	3.6	1.7	0.08	13.6	0.5	<1
Experimental (first run Foxy[a])	<0.5	–	0.5	1.5–2	1	
Lowest grade of commercial polycrystalline silicon	<0.1	<0.1	<0.1	<0.1		<4
MG-Si (Silgrain, Elkem)	400	21	1100	25–35	25–35	<50

Note: For comparison, the impurity content of low-grade electronic silicon and of high-purity metallurgical silicon (Silgrain from Elkem) is also shown.

[a]   Ingot FS1, average values up to 60% of the growth height [49].

Nevertheless, P gettering was demonstrated very effective in enhancing the lifetime by a factor of 100, bringing the substrate to solar quality. This result is in good agreement with the arguments reported in Chapter 1 of this book, which show the need of accomplishing for the entire silicon value chain in order to define the PV quality of a silicon feedstock.

2.7   CONCLUSIONS

Both the outcomes of the earlier projects and of the most recent ones seem to show that there is not an apparent advantage in the use of high-purity raw materials in view of the optimization of the MG-silicon quality (mostly concerning its B- and

P-concentration), although respectable solar cell efficiencies were systematically obtained from these feedstocks.

The potential benefits concerning the dopant contamination find a negative counterpart in difficult furnace conduction, that eventually leads to high current consumption, high loss of SiO, and consequent modest quartz conversion ratios, as compared with standard furnace operations.

Thermodynamics is eventually against the production of high-quality silicon from the carbothermic reduction of quartz, as high process temperature would inevitably lead to any sort of impurity contamination, almost independently of the grade of the raw materials and of the furnace construction materials, which tend to deteriorate with time.[*]

The cost of high-purity quartz is the last, but not the least problem, that could be however overcome in the case of its future massive use in standard furnace operation. The use of high-purity synthetic silica would be an alternative solution.

The use of low boron quartz in standard furnace operations[†] would, in any case, lead to several advantages, which will be fully apparent in Chapter 3, considering the poor B-removal efficiency of chemical and MG processes and the possibility to drop these processes from the whole purification chain.

REFERENCES

1. A. Schei, J. K. Tuset, and H. Tveit, 1988. *Production of High Silicon Alloys*, Tapir, Trondheim.
2. E. J. Øvrelid, K. Tang, Th. Engh, M. Tangstad, 2009. Feedstock, in K. Nakajima and N. Usami (Eds.), *Crystal Growth of Silicon for Solar Cells*, Springer, Berlin.
3. J. O. Odden, G. Halvorsen, H. Rong, R. Glockner, 2008. Comparison of the energy consumption in different production processes for solar grade silicon, in *Silicon for the Chemical and Solar Industry IX*, Oslo, Norway, June 23–26, 2008, pp. 1–16.
4. B. Andersen, 2010. The metallurgical silicon process revisited, in *Silicon for the Chemical and Solar Industry X*, Alesund-Geiranger, Norway, June 28–July 02, pp. 11–23.
5. B. Ceccaroli, S. Pizzini, 2012. Processes, in S. Pizzini (Ed.), *Advanced Silicon Materials for Photovoltaic Applications*, Wiley, Chichester, UK, pp. 21–78.
6. A. Agarwal, 1988. Mechanism in Tuyere injected reactors for the carbothermic reduction of silica, Master thesis, MIT.
7. A. Schei, J. Kr. Tuset, H. Tveit, 1998. *Production of High Silicon Alloys*, Tapir, Forlag, pp. 29.
8. P. J. Guichelaar, 1997. *Acheson Process*, Springer-Verlag.
9. B. S. Xakalashe, M. Tangstad, 2011. Silicon processing: From quartz to crystalline solar cells, in *Proceedings of Southern African Pyrometallurgy*. R. T. Jones and P. den Hoed (Eds.), *Southern African Institute of Mining and Metallurgy*, Johannesburg, pp. 83–99.
10. L. P. Hunt, V. D. Dosaj, J. R. McCormick, L. D. Crossman, 1976. Production of solar grade silicon from purified MG silicon, *Record of the 12th IEEE Photovoltaic Specialist Conference*, Baton Rouge, IEEE.org (USA), pp. 125.

---

[*] S. Pizzini, unpublished results.
[†] Carried out, however, in the cleanest conditions possible.

11. L. P. Hunt, V. d. Dosaj, J. R. McCormick, Solar silicon via improved and expanded metallurgical silicon technology, *JPL Contract no. 954559, ERDA/JPL 954559, 76/1-2, 77/1-3, 78/1-7).*

12. L. P. Hunt, V. D. Dosaj, 1979. Progress on the Dow Corning process for solar grade silicon, *Proceedings of the 2nd European Photovoltaic Conference*, pp. 98–105.

13. J. A. Amick, J. P. Dismukes, R. W. Francis, L. P. Hunt, P. S. Ravishankar, M. Schneider, K. Matthei, R. Sylvain, K. Larsen, A. Schei, 1985. Improved high purity arc furnace silicon for solar cells. *J. Electrochem. Soc.*, 132, 339–345.

14. R. Davis, R. H. Hopkins, A. Rohatgi, H. H. Hines, P. Rai-Chaudury, H. C. Hollenkopf, 1982. Effect of impurities and processing on solar silicon solar cells, *Final Report*, JPL no. 9850, Contract 954331, February 1982, pp. 1–226.

15. S. Pizzini, C. Calligarich, 1984. On the effect of impurities on the photovoltaic behavior of solar grade silicon. I. The role of boron and phosphorous primary impurities in p-type single crystal silicon *J. Electrochem. Soc.*, 131, 2128–2132.

16. S. Pizzini, L. Bigoni, M. Beghi, C. Chemelli, S. Fossati, M. Tincani, 1986. On the effect of impurities on the photovoltaic behavior of solar grade silicon. II. The influence of titanium, vanadium, chromium, iron and zirconium on PV behaviour of polycrystalline solar cells. *J. Electrochem. Soc.*, 133, 2363–2373.

17. M. Rodot, J. E. Burree, A. Mesli, G. Revel, R. Kishore, S. Pizzini, 1987. Al-related recombination centres in polycrystalline silicon. *J. Appl. Phys.*, 62, 2556–2558.

18. G. L Coletti, D. MacDonald, D. Yang, 2012. Role of impurities in solar silicon in advanced silicon materials for photovoltaic applications, in S. Pizzini (Ed.), *Advanced Silicon Materials for Photovoltaic Applications*, Wiley, Chichester, UK, pp. 79–125.

19. S. Pizzini, 2015. Growth of semiconductor materials, in *Physical Chemistry of Semiconductor Materials and Processes*, Wiley, Chichester, UK, pp. 290–294.

20. S. Pizzini, C. Chemelli, M. Gasparini, M. Rustioni, 1984. Italian Patent 203501.

21. B. Andresen, 2006. The silicon process, in *Silicon for the Chemical and Solar Industry VIII*, Trondheim, Norway, June 12–15, pp. 35–50.

22. G. Halvorsen, 1985. Method for production of pure silicon, US Patent 4,539,194.

23. E. Schurrmann, H, Litterscheid, and P. Funders, 1974. Investigation of the melting equilibria of the phase diagram calcium-silicon. *Arch. Eisenhuttenwes.*, 45, 367–371.

24. H. Sigmund, 1982. Solubilities of magnesium and calcium in silicon. *J. Eletrochem. Soc.*, 129, 2809–2812.

25. D. P. Fenning, B. K. Newman, M. I. Bertoni, S. Hudelson, S. Bernardis, M. A. Marcus, S. C. Fakra, T. Buonassisi, 2013. Local melting in silicon driven by retrograde solubility. *Acta Mater.*, 61, 4320–4328.

26. Y. Shen, 2015. Carbothermal synthesis of metal-functionalized nanostructures for energy and environmental applications. *J. Mater. Chem. A*, 3, 13114–13118.

27. M. Rustioni, D. Margadonna, R. Pirazzi, S. Pizzini, 1986. Solar silicon from directional solidification of MG silicon produced via the silicon carbide route. *JPL Proceedings of the Flat-Plate Solar Array Project Workshop on Low-Cost Polysilicon for Terrestrial Photovoltaic Solar-Cell Applications*, pp. 297–321 (SEE N86–26679 17–44).

28. H. A. Aulich, F. W. Schulze, J. G. Grabmaier, 1984. Verfahren zur Herstellung von Solarsilicium. *Chemie Ingenieur Technik*, 56, 667–673.

29. H. A. Aulich, F. W. Schulze, H. P. Urbach, and A. Lerchenberger, 1986. Solar-grade silicon prepared by carbothermic reduction of silica, in *JPL Proceedings of the Flat-Plate Solar Array Project Workshop on Low-Cost Polysilicon for Terrestrial Photovoltaic Solar-Cell Applications*, pp. 267–275.

30. J. Grabmaier, 1986. *Siemens Forschungsund Entwicklungsberichten*, Vol. 15, pp. 157–162.

31. H. A. Aulich, K. H. Eisenrith, F. W. Schulze, B. Strake, H. P. Urbach, 1986. Assessment of advanced carbothermic reduction process for production of high-purity silicon. *Proceedings of the Electrochemical Society Meeting*, Boston, MA, May 4–9, pp. 443–451.

32. H. A. Aulich, K. H. Eisenrith, F. W Schulze, H. P. Urbach, 1986. Solar-grade silicon prepared by carbothermic reduction of silica. *PL Proceedings of the Flat-Plate Solar Array Project Workshop on Low-Cost Polysilicon for Terrestrial Photovoltaic Solar-Cell Applications.* pp. 267–275 (SEE N86-26679 17-44): 1986fpsa.proc.267A.

33. H. A. Aulich, K. H. Eisenrith, F. W Schulze, A. Lerchenberg, and H. P. Urbach, 1987. Removal of carbon and SiC from silicon prepared by advanced carbothermic reduction. *Proceedings of the 7th ECPV Conference,* Sevilla, Spain, pp. 731–735.

34. H. A. Aulich, F. W. Schulze, High-purity lining for an electric low shaft furnace, US patent 4971772 A.

35. F. W. Schulze, H. J. Fenzl, K. Geim, H.-D. Hecht, H. A. Aulich, 1984. *Proceedings of the 17th IEEE Photovoltaic Specialists Conference*, May 1984, Kissimee, USA, IEEE. org, p. 584.

36. T. Noda, 1985. Move to development of low-cost silicon in Japan, *Proceedings Flat-Plate Solar Array Project*, DOE/JPL 1012–122, pp. 213–231.

37. S. Amendola, Method for making silicon for solar cells and other applications. WO/200/106860.

38. H. A. Aulich, F. W. Schulze, H. P. Urbach, A. Lerchenberger, 1985. Solar grade silicon prepared by carbothermic reduction of silica. *JPL Proceedings of Flat-Plate Solar Array Workshop on Low Cost Polysilicon for Terrestrial Photovoltaic Solar Cells,* October 28–30, 1985, Las Vegas, pp. 267–275.

39. T. Naoshima, A. Yamashita, C. Ouchi, Y. Oguchi, 2002. Concentration and behavior of carbon in semiconductor silicon. *J. Electrochem. Soc.* 117, 1566–1568.

40. S. De Wolf, J. Szlufcik, Y. Delannoy, I. Périchaud, C. Häßler, R. Einhaus, 2002. Solar cells from upgraded metallurgical grade (UMG) and plasma-purified UMG multicrystalline silicon substrates. *Solar Energy Mater. Solar Cells*, 72, 49–58.

41. L. Ottem, 1993. Løselighet og termodynamiske data for oksygen og karbon i flytende legeringer av silisium og ferrosilisium, SINTEF Report STF34 F93027.

42. R. T. Dolloff, 1960. WADD Report 60–143.

43. R. W. Olesinski, G. J. Abbaschian, 1984. *C-Si Bulletin of Alloy Phase Diagram*, 5, 486.

44. R. I. Scace, G. A. Slack, 1959. Solubility of carbon in silicon and germanium, *J. Chem. Phys.*, 30, 1551–1555.

45. E. Øvrelid, B. Geerligs, A. Wærnes, O. Raaness, I. Solheim, R. Jensen, K. Tang, S. Santeen, B. Wiersma, 2006. Solar grade silicon by a direct metallurgical process, *Silicon for the Chemical Industry VIII*, 223–239.

46. L. J. Geerligs, P. Manshanden, G. P. Wyers, E. J. Øvrelid, O. S. Raaness, A. N. Waernes, B. Wiersma, 2005. Specifications of solar grade silicon: How common impurities affect the cell efficiency of MC-Si solar cells, *Presented at the 20th PVSEC*, Barcelona, Spain.

47. M. DiSabatino, E. J. Overlied, R. Kpecek, S. Binetti, V. D. Mihailetchi, L. Geerligs, A. N. Vaernes, 2009. Foxy development of solar grade silicon feedstock for crystalline wafers and cells by purification and crystallization, *Proceedings of the 24th Photovoltaic Solar Energy Conference*, Hamburg, Germany, pp. 1823–1826.

3 Conventional and Advanced Purification Processes of MG Silicon

Yves Delannoy, Matthias Heuer,
Eivind Øvrelid, and Sergio Pizzini

CONTENTS

3.1   INTRODUCTION

Studies and developments of processes which could be used to convert metallurgical grade (MG)-silicon or any kind of impure silicon in a solar grade (SoG) feedstock started early in the 1980s in the United States, Europe, and Japan with almost common objectives and strategies [1].

The main aim was to demonstrate that physical, chemical, and metallurgical treatments of MG-Si in the molten or solid state, would bring it to a SoG quality, and then to photovoltaic (PV) applications. These processes would avoid its full conversion in a gaseous compound (trichlorosilane or silane), as is the case of the Siemens or of the fluidized-bed processes, which were at that time too expensive to retard PV development dramatically.

In spite of the revolutionary improvement of gas phase processes and the consequent decrease of the cost of electronic grade (EG) silicon, which occurred in the last two decades, R&D in the field of MG-Si upgrading continued and still continues, with the aim to develop a SoG process capable to compete in terms of cost and environmental friendship with the gas phase processes [2–11].

Most of these studies were very successful in terms of basic results, few in terms of industrial applications, with the unique exception[*] of the Elkem process (see Chapter 4).

The main result was the identification and the preliminary development of a sequence of basic steps addressed at the selective removal of impurities [6].

This sequence consists of

- A stage based on vacuum sublimation, carried out mostly to remove phosphorous
- A stage based on the use of reactive gases (oxygen and chlorine) bubbled in a molten Si bath to remove volatile impurities such as chlorides and oxides
- An intermediate stage based on the use of liquid/liquid impurity extraction processes using molten salts or metals as extracting media
- A hydrometallurgical process on silicon powders, obtained by crashing the material upgraded in the first three steps, addressed at the solubilization of metallic impurities with acid solutions, mostly based on HCl
- And a final purification stage based on the directional solidification (DS) of a molten silicon bath to get an oriented polycrystalline ingot of SoG quality

The benefits of the first two steps, which could be carried out either in a vacuum chamber (the first one) or in a furnace with appropriate gas manifolds, are associated to the high vapor pressure of P and to the volatility of many metallic chlorides ($FeCl_3$ and $AlCl_3$) and of some oxides, as boron oxide (BO), at the melting temperature of silicon. The volatility of BO could be increased using wet gases, which favor the formation of boron hydroxides.

The intermediate step is a true pyrometallurgical process used for the liquid/liquid extraction of metallic and nonmetallic impurities which present favorable

---

[*] Silicor and Ferroatlantica are recent additional exceptions, but not yet at an industrial stage, see later in this chapter.

segregation coefficients in suitable slags. Silicates, carbonates, and fluorides are the most common extracting media for silicon.

At the end of the slagging process, liquid silicon could be mechanically separated from the reacted slag if the respective densities are different, or the mixture is cooled down at room temperature and the silicate matrix is dissolved in an acid solution, after grinding, in the due of a hydrometallurgical process.

The success of the hydrometallurgical process depends on the size of the Si powders used, as the HCl-based acid solutions[*] interact only with impurities segregated[†] at the surface of the grains and the yield of the process of repartition of impurities between the bulk and the surface depends on the surface area.

The yield and the safety of this process is limited by the pyrophoricity of the silicon powders in air and in aqueous media, while it is favored by the use of suitable alloys of silicon, as is the case of Ca–Si alloys, which will be discussed in Section 3.2.4.

As an alternative to the use of slags as extracting media, metals might also be used, with Al as the most favorable candidate. In this case, purified silicon is separated mechanically from the reaction mixture and then submitted to a further purification using DS to grow a silicon ingot. We will show that this process is now carried out at industrial level by Silicor (see Section 3.3.2).

Eventually, a DS growth provides the last purification process, arising from the repartition of the impurities between the liquid and solid phase, favored by the very low segregation coefficients of most metallic impurities in silicon. The limit of this process, with respect to PV applications, is a practically negligible B and P segregation efficiency as we have seen in Chapters 1 and 2.

Each process step presents a number of challenges, most, but not all, of them have been practically overcome in the course of more than four decades of research and industrial development, within the limits dictated either by chemistry or physics.

Silicon compatibility with crucible materials adds additional limits in terms of process temperatures and contamination [12]. Typical, but not exhaustive, is the case of P-removal by vacuum evaporation (see Section 3.2.1), where the use of graphite containers is needed to operate the process at temperatures well above the maximum temperature of quartz crucibles utilization, with the drawback of heavy carbon contamination.

Due to massive R&D investments carried out in the last 10 years to optimize the traditional polysilicon routes,[‡] which lead to very large cost reductions, it could be now questioned whether advanced purification of MG-Si could compete in terms of cost (and PV conversion efficiency) with advanced gas phase processes.

To arrive at a conclusion, we intend to discuss first the basic chemical and physical backgrounds of the processes envisaged for this aim and then to report on some relevant processes which are close to a full industrial stage. A full chapter (Chapter 4) will be, however, devoted to the process developed by Elkem, in view of its already being in the market.

---

[*] Silicon is insoluble in HCl solutions.
[†] As individual impurities or as separate phases in the case of supersaturation conditions.
[‡] Siemens and fluidized bed reactor (FBR).

3.2 CHEMISTRY AND PHYSICS OF MG-Si PURIFICATION

3.2.1 VACUUM SUBLIMATION

As already shown in Chapter 2, some of the impurities present in MG-Si (P, Mg, Al, and Ca) exhibit relatively high vapor pressures at the melting temperature of silicon (see Figure 3.1) [13].

Their actual vapor pressures depend, of course, on their concentration according to the equation

$$p_i = p_i^o x_i\, f(x_i) \tag{3.1}$$

which holds for single impurity melts.

In the ideal solution approximation, holding for very dilute solutions, the $f(x_i)$ term is close to one and the values reported in Figure 3.1 should be lowered by a factor 10^3–10^6, depending on the specific impurity concentration x_i, to get the final theoretical pressure.

If strong interaction among a dissolved impurity and the silicon solvent occurs, the thermodynamics of the vapor/liquid equilibrium depends on the details of the

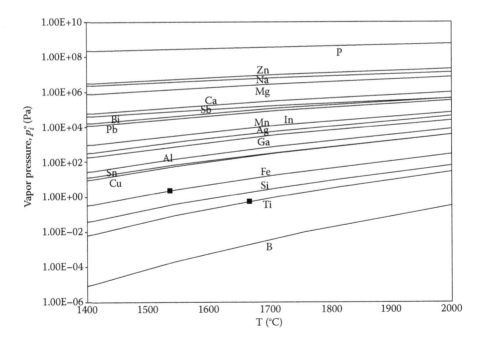

FIGURE 3.1 Vapor pressures of the elements as a function of the temperature. (After J. Safarian and M. Tangstad, 2012. Vacuum refining of molten silicon, *Metall. Mater. Trans. B,* 43 1427–1445. Open access article distributed under the terms of the Creative Commons Attribution License, which permits unrestricted use, distribution, and reproduction in any medium, provided the original work is properly cited.)

interaction process, which could be described by several models, starting from the regular solutions one [12, pp. 29–33]. The kinetics of impurities evaporation is, however, the main factor for the practical feasibility of the process, which implies at least a diffusive transfer of the impurity from the bulk of the sample to the impurity-depleted surface and a fast evaporation process from the melt surface.

Vacuum sublimation might be, therefore, enhanced either by mechanical stirring or by local heating of a thin surface layer of molten silicon, where surface renewal might be very fast, using plasma or e-beam heating.

Among the different impurities contaminating MG-Si, which has been dealt with in Chapter 1, Figure 3.1 shows that P (but also Al, Ca, and Mg) does present the most favorable conditions to be removed by vacuum evaporation from a molten MG-Si bath, with a limited loss of silicon. This was the main reason why vacuum evaporation was considered, since the end of the 1980s, a promising route for volatile impurities removal from MG-Si.

As an example, the removal of P, Ca, and Al impurities from molten silicon by vacuum evaporation has been originally studied by Suzuki et al. [14] starting with a MG-Si sample containing a total amount of about 1000 ppmw of heavy elements and 10 ppmw of B and P.

They showed that the rate of removal depends on the nature of the chemical species, increases with the melt temperature and decreases with time (see Figure 3.2), and that Ca is together with P the impurity easier to be removed.

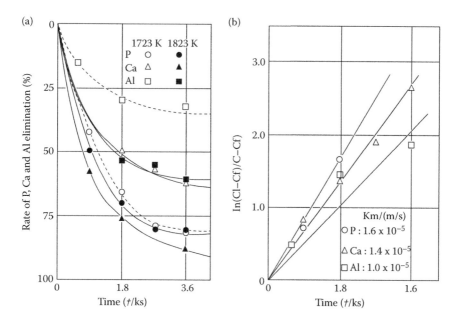

FIGURE 3.2 (a) Rate of removal of P, Ca, and Al from molten silicon at 0.027 Pa. (b) First-order reaction kinetics for P, Ca, and Al evaporation impurity concentration in atoms. (After K. Suzuki et al., 1990. *J. Jpn. Inst. Met.*, 54, 161–167. With permission of the Japan Institute of Metals and Materials.)

They also show that the evaporation rate follows approximately a first-order kinetics

$$-\frac{dC}{dt} = k_m \frac{(A/V)}{(C - C_f)} \tag{3.2}$$

$$\ln \frac{C_i - C_f}{(C - C_f)} = k_m (A/V)t \tag{3.3}$$

(see Figure 3.2b), where C_i is the impurity concentration, C_f is its equilibrium concentration, k_m is the mass transfer coefficient, and A and V are the surface area and volume of the melt, respectively.

They eventually demonstrated that the rate determining step of the evaporation process of these impurities is their diffusion in the bulk toward the melt surface, once the initial impurity content has been removed.

The issue has been recently revisited by S. Zheng et al. [15], who were able to show (see Figure 3.3) that at relatively low temperatures ($T < 2173$ K), the mass transfer coefficient of the process K_p fits with that of the surface evaporation K_c, while at higher temperatures, the rate determining step is the mass diffusion in the melt, with a rate constant K_m.

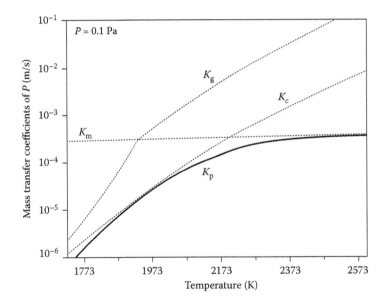

FIGURE 3.3 Calculated temperature dependence of the mass transfer coefficients: K_p is the effective mass transfer coefficient while K_c accounts for surface evaporation, K_g for the mass transfer in the gas phase, and K_m for the mass transfer in the melt. (With kind permission from Springer Science+Business Media: *Metall. Mater. Trans. A*, Numerical simulations of phosphorous removal from silicon by induction vacuum refining, 42a, 2011, 2214–2225, S. Zheng, T. A. Engh, M. Tangstad, and X. I. Luo.)

A. Souto et al. [16] applied the vacuum refining process to an already purified MG-Si material, with an initial Ca and P concentration of 1.4 and 3.3 ppmw. The process was carried out in an electrically heated furnace, held at 0.1 Pa during the evaporation process, using a graphite container for a charge of silicon of 500 kg.

Working in the temperature range 1550–1650°C, variable concentrations of Ca and P could be evaporated from the charge, with amounts significantly increasing with the increase of temperature and process time (see Figure 3.4).

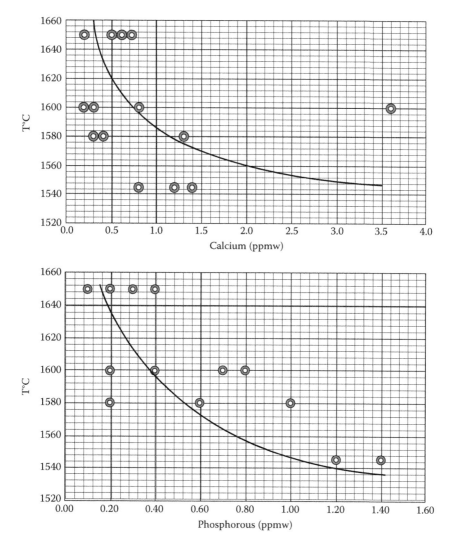

FIGURE 3.4   Temperature dependence of a vacuum evaporation process on the P and Ca content of molten charges of 500 kg of silicon. The sequence of isothermal concentration in the diagrams corresponds to increasing process times in the range of 180 and 510 min. (Data after A. Souto et al., 2014. Industrial scale vacuum applications in the FerroSolar project, in *Silicon for the Chemical and Solar Industry XII*, Trondheim, Norway, pp. 67–76, June 23–26.)

While it is demonstrated that the process works well in the case of P, some spread of data at intermediate temperatures is observed in the case of Ca, whose origin could have the same nature (previous oxidation treatments) as for the case of crystal system experiments carried out within a Department of Energy (DOE) project in the United States [17].

They conducted sequential vacuum evaporation tests at temperatures around 1500°C,[*] starting from a purified commercial MG-Si. Figure 3.5 displays the results of a representative experiment carried out at 1500°C, starting from a sample containing 8 ppm of Al, 15 ppm of P, and 30 ppm of Ca. The mass of the samples amounted to 17 kg. It could be seen that the rate of the evaporation process depends on the nature of the impurity, only P showing a systematic decrease with the increase of the total process time, while the concentration of Ca and Al remains unaffected in the first runs, followed by a rapid decrease at the end of the experiment.[†]

In the course of the same experiment, it was also observed that the half time of the process increases with the decrease of the impurity content and was of the order of 15 h in the last run, enabling a final P-concentration of 0.3 ppmw to be reached.

The kinetics of P volatilization might be understood considering that the rate of the process is melt-diffusion controlled and that the composition of the melt affects the yield of the process, due to impurity interaction events, with a consequent decrease of the *isolated* impurities concentration in solution and of their vapor

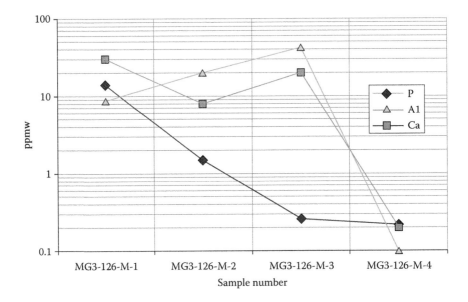

FIGURE 3.5 Kinetics of P, Al, and Ca vacuum evaporation at 1500°C. Each run after the first one starts with a portion of the material already vacuum treated. (Data after C. P. Khattak, D. B. Joyce, and F. Schmid, 2008. Development of solar grade (SoG) silicon. *Final Report for DOE SBIR Phase II, Contract Number DE-FG02-04ER83928,* January 16. Courtesy of C. P. Khattak.)

---

[*] Compatible with the use of quartz crucibles.
[†] No explanation is given by the authors for the anomalous behavior of Ca and Al.

pressure and diffusivity. This conclusion is in good agreement with the results of Safarian and Tangstad [18], who showed that P removal is faster from cleaner melts.

Also, the anomalous behavior of Ca and Al in Figure 3.5 might be understood as due to impurity interaction processes.

In fact, if the MG-Si sample used in the experiments would come from a feedstock originally submitted to an oxidizing treatment (see next section), Ca and Mg would be present in the form of oxide (micro-)precipitates* which remain distributed in the solid matrix after cooling.

Ca and Al present in a MG-Si feedstock in the form of oxides cannot be removed by a vacuum treatment, but could, instead, segregate at the sample surface during a long thermal anneal, simulating the effect of a vacuum evaporation.

Among the number of vacuum evaporation attempts, P removal by local e-beam heating of a layer of molten MG-Si continuously fed into a water-cooled copper mold has been successfully proven by N. Yuge et al. [2] in the frame of a Japanese New Energy and Industrial Technology Development Organization (NEDO) project involving the company Kawasaki Steel.†

It was demonstrated that P content could be decreased to 0.1 ppmw, depending, however, on the silicon supply rate (see Figure 3.6), but no information is given about the contemporary loss of silicon.

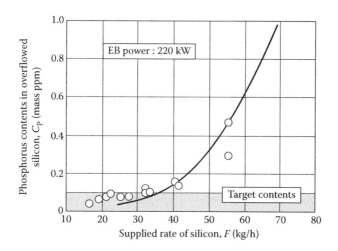

FIGURE 3.6 Dependence of the P-content on the silicon supply rate. (After N. Yuge et al.: Hiwasa, and F. Aratani. Purification of metallurgical-grade silicon up to solar grade. *Prog. Photovolt. Res. Appl.*, 9, 203–209. 2001. Copyright Wiley-VCH Verlag GmbH & Co. KGaA. Reproduced with permission.)

---

* Whose presence could not be revealed by scanning ion mass spectrometry (SIMS) or other analytical techniques.
† In this process, the P concentration is initially reduced under vacuum, followed by a first directional solidification (DS) step to reduce the concentration of Al and Fe impurities. Boron is then removed from the surface by reaction with Ar plasma and water vapor, and finally a second DS growth leads to a good SoG silicon. Laboratory results were encouraging, but the overall cost of the process was presumably too high, also due to the use of two DS growths.

An unavoidable drawback of vacuum evaporation processes is the condensation of impurity vapors on the cold zones of the vacuum system[*] and to impurity contamination (e.g., carbon or oxygen) arising from the melt-container walls interaction. The latter could be, however, minimized (but never suppressed) by a suitable choice of container materials.

For all these reasons, vacuum sublimation of P from MG-Si melts is still considered a promising technological step for the production of SoG-Si [7], yet some concerns remain for its full industrial utilization.

3.2.2   Gas Phase Purification (Dry Gases)

Gas (O_2 and Cl_2) blowing in a ladle filled with molten silicon has been, and still is, an industrial practice in the MG-Si industry, used to reduce the concentration of impurities which could be selectively oxidized or chlorinated. Impurity removal by blowing dry and wet oxygen has also been a widely studied process used to upgrade MG-Si to solar silicon [6,19–21]. The theoretical yield of these processes depends on the thermodynamic equilibrium relationships between dissolved impurities and the corresponding oxygenated

$$m\text{Me} + n\text{O}_2 \rightleftharpoons \text{Me}_m\text{O}_{2n} \tag{3.4}$$

or chlorinated species

$$m\text{Me} + n\text{Cl}_2 \rightleftharpoons \text{Me}_m\text{Cl}_{2n} \tag{3.5}$$

or hydrated species, in the case of use of wet gases (see next section).

Chlorination processes are particularly appropriate for Al and Fe removal, due the volatility of the corresponding chlorides, but hard environmental and materials degradation phenomena occur[†] at temperatures above the melting temperature of silicon using chlorine gas, discouraging their use in SoG-Si production processes.

Milder conditions could be, however, applied using silicon tetrachloride ($SiCl_4$) instead of chlorine as the chlorinating gas, to remove Al as volatile $AlCl_3$ from molten MG-Si [22]. Working at 1450°C on a molten silicon bath with an initial Al concentration of 1527 ppm, a residual concentration of 128 ppm has been measured after 60 min of treatment. The calculated effect of a multiple step procedure shows that after four subsequent steps the final Al content would reach a value close to 0.5 ppmw.

The results are encouraging, and of potential use[‡] as the final purification of a silicon feedstock purified with the Al-smelting process (see Section 3.2.6).

Oxidation processes present, instead, the best practical conditions for a preliminary purification process of a MG-Si meltstock from Ca, Mg, Al, and Ti impurities,

---

[*] With the formation of crusts that could fall back in the silicon melts.
[†] But also the volatilization of silicon in the form of silicon chlorides.
[‡] This is the opinion of one (S.P.) of the chapter's author, not of the company.

as could be observed in the Ellingham diagram reported in Figure 3.7 [23]. It does allow, in fact, the identification of these impurities as those that present the thermo-dynamic conditions to be preferentially oxidized and which, therefore, would allow favorable process paths.

From Figure 3.7, one could see, in fact, that the oxidation of Ti, Al, Mg, and Ca is thermodynamically favored with respect of that of silicon. These impurities are, therefore, the possible targets of a preferential oxidation process and their residual (theoretical) concentration might be calculated for the equilibrium between a dis-solved impurity Me, a divalent oxide, and silica in a silicon melt (with silicon, MeO, and SiO_2 at unit activity) by writing the following equation:

$$2Me(a_{Me}) + SiO_2 \rightleftharpoons 2MeO + Si \qquad (3.6)$$

where a_{Me} is the residual Me activity in equilibrium conditions, with SiO_2, MeO, and Si at unitary activity.

In turn

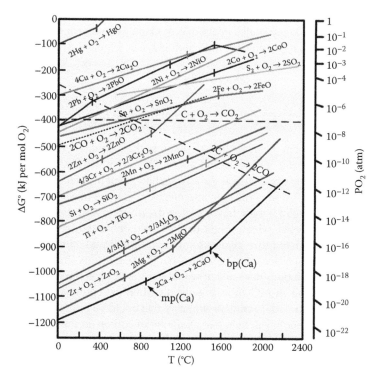

FIGURE 3.7 Standard Gibbs free energies for a set of metal oxidation reactions. (After Y. Shen, 2015. Carbothermal synthesis of metal-functionalized nanostructures for energy and environmental applications, *J. Mater. Chem. A*, 3, 13114–13118. Reproduced by permission of The Royal Society of Chemistry.)

$$a_{\text{Me}} = \exp{-\frac{2\Delta G_{\text{MeO}} - \Delta G_{\text{SiO}_2}}{2RT}} \qquad (3.7)$$

where ΔG^o_{MeO} is the standard Gibbs free energy of MeO formation, $\Delta G^o_{\text{SiO}_2}$ is the corresponding value for silica at the melting temperature of silicon, and the activity $a_{\text{Me}} - \gamma_{\text{Me}}x_{\text{Me}}$ would be equivalent to the impurity concentration x_{Me} in the range of dilute solutions, for which the activity coefficient γ_{Me} is unitary.

The calculated, residual absolute activities of Ca, Mg, Ti, and Al at 1763 K amount, respectively, to 10^{-4}, 10^{-2}, $10^{-0.5}$, and 10^{-3}, showing that dry oxidation is particularly effective for Ca and Al. These figures are in good agreement with those reported by Tuset [24], who quotes a concentration of 8.06×10^{-2} (Al w%) and 9.94×10^{-3} (Ca w%) at 1550°C.

The oxidation with oxygen or with mixtures of oxygen in an inert gas* is a practical tool, but suffers from unavoidable, silicon oxidation losses, which, however, favor the formation of a supernatant silicate slag as the stable final oxidation product in which impurities are collected.

As the equilibrium oxygen pressure in Si that arises when silicon is in equilibrium with its oxide is that needed to oxidize Ca, Mg, Ti, and Al, the oxidation of these impurities could be carried out by equilibrating a MG-Si melt with SiO_2 [25]. An alternative to SiO_2, $SiO^†$ could also be used as oxidant. The amorphous SiO phase is stable, but it easily disproportionates to Si and SiO_2 (see Figure 3.8)

$$2SiO \rightleftharpoons SiO_2 + Si \qquad (3.8)$$

as the Gibbs free energy of reaction (3.8) is slightly more negative than that of SiO formation [26].

Both processes present the advantage to suppress the loss of silicon by side oxidation but have the drawback of kinetic hindrances. The use of silica powder implies, in fact, the oxidation of the impurities at the silica/silicon melt interface,‡ a process that involves their slow diffusion at the reaction interface. The use of SiO vapors implies the direct interaction of SiO§ with the impurities dissolved in the liquid silicon phase, a potentially slow process, which might be, however, enhanced by blowing SiO in the silicon mass associated to a vigorous stirring of the liquid bath. Details on the process kinetics are not reported in literature, but the best results should be obtained with the use of an induction furnace that would enable electromagnetic stirring.

Also carbon, present in MG-Si as dissolved carbon or segregated as SiC, when the melt is carbon saturated, is another impurity that could be removed by oxygen.

One can see in Figure 3.7, in fact, that oxidation of carbon (at unit activity) to CO is thermodynamically favored with respect to Si oxidation at temperatures slightly higher than 1500°C.

---

* Not nitrogen, as nitrogen will nitridate silicon.
† SiO is a commercial product obtained by condensing SiO vapors.
‡ And the loss of crucible material.
§ And SiO_2 as its decomposition product.

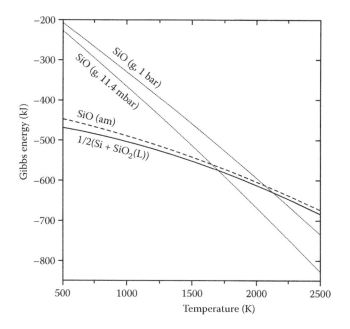

FIGURE 3.8   Temperature dependence of the Gibbs free energies for the SiO formation (dotted curve) and its decomposition. (Reprinted from *J. Non-Cryst. Sol.*, 336, S. M. Schnurre, J. Grobner, and R. Schmid-Fetzer, Thermodynamics and phase stability in the Si–O system, 1–25, Copyright 2004, with permission from Elsevier.)

Decarburization of silicon melts is particularly important for solar silicon melts produced by reaction of high purity silica with high purity carbon (see Chapter 2), as these melts are carbon saturated and DS is only a partial solution [27]. The oxidative decarburization of MG-Si has been experimentally studied, among others, by Sakaguchi and Maeda [28].

They used as the oxygen source either the quartz crucible in which a molten silicon sample* was kept under reduced pressure (10^{-3} atm) or silica powder added to the melt.

The results obtained with the first method are reported in Figure 3.9 that shows that by simply melting a silicon sample contaminated with 1200 ppmw of C in a quartz crucible, decarburization occurs up to 100 ppmw. C contents lower than 1 ppmw were obtained by adding silica powder. Removal of CO under low vacuum is necessary to avoid reaction reversion.

More recently, Søiland [29] carried out a careful reexamination of the theoretical background of the carbon oxidation process, putting in evidence that the difficulty of CO exhaustion could be the reason of modest results obtained with large furnaces, as also in the case of plasma treatments.

Decarburization of MG-Si was also carried out using SiO as the oxidant [30] immediately after the tapping process.

---

* The experiments were carried out with small samples to facilitate the reaction.

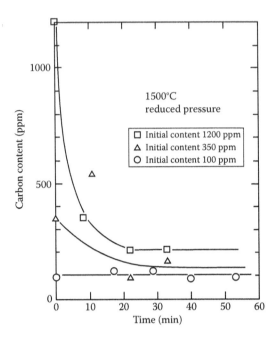

FIGURE 3.9 Carbon content as a function of time when a molten silicon sample is kept in contact with a silica crucible. (With kind permission from Springer Science+Business Media: *Metall. Trans. B*, Decarburization of silicon melt for solar cells by filtration and oxidation, 23B, 1992, 423–427, A. K. Sakaguchi and M. Maeda, October 2, 2015.)

The results are particularly interesting, as could be seen in Table 3.1, that shows that a relatively short test (48 min) carried out at 1700°C allows bringing the final C content at 5 ppmw in a MG-Si sample with an initial content of 1120 ppmw.

3.2.3 BORON REMOVAL BY WET GAS OR PLASMA

A purely oxidative refining has little or no effect on B-concentration, due to the unfavorable thermodynamics of the process, in spite of the volatility of BO, that shifts to the right (Equation 3.4).

The process of B-removal could be, instead, carried out using wet gases (mixtures of hydrogen or inert gases with water vapor) [20,31–35] that favor the volatilization of B as boron hydride or oxyhydride species.

It is, however, known [36,37] that the reaction does not occur at a reasonable rate unless the partial pressure of oxygen is maintained lower than that leading to the formation of a surface layer of SiO_2, that would block the interface reaction.

B removal might be better carried out in the presence of steam or/and hydrogen, or operating the process under plasma conditions, which favor the dissociation of water vapor

$$H_2O \rightleftharpoons H_2 + \tfrac{1}{2}O_2 \tag{3.9}$$

TABLE 3.1

Effect of Silicon Decarburization with SiO

Total time (min)	0	6	36	39	42	45	48
Blowing time (min)	0	6	6	9	12	15	18
Hold time (min)	0	0	30	30	30	30	30
Carbon content (ppmw)	1120	580	576	117	36	13	5

Source: Data from J. Hintermayer, 2011. Process for decarburization of a silicon melt WO 2011088953 A1 28/07/2011 assigned to Evonik, Degussa.

with the formation of ionized and radicalic species that are kinetically very active, as shown by early research using inductive plasma in France [37,38] and arc plasma in Japan [36,39,40].

The NEDO process (see Chapter 6) and the Photosil process (see Section 3.3.1.2) are examples of processes developed on the industrial scale in Japan and in France, respectively,* but the need remains to improve the boron removal yield and rate for a full accomplishment with PV requests.

Understanding the kinetic aspects and increasing the rate of the B-purification processes by wet and plasma processes are, therefore, still hot research subjects not only in France [41,42], but also in Norway [43,44], China [19], Korea [45], and the United States [46].

Wet and plasma processes might differ by

- The type of injection: Bubbling into the liquid (see Figure 3.10) or blowing at the surface (Figure 3.11)
- The mixture of injected: Plasma or "cold" gas
- The nature of reactive gases: O_2 and H_2, or H_2O and eventually H_2

However, both have in common an interface between the liquid silicon and a gaseous phase, where the boron removal takes place. In a plasma process, the reactive species can be provided as H and O atoms, or $OH^•$ radicals, but the stable reacting species at the temperature of liquid silicon is always H_2O. The thermodynamics of the process can thus be discussed considering H_2O as an oxidant, keeping in mind that the actual kinetic behavior could differ, depending on the process route (wet or plasma).

Boron is removed in such processes in the form of gaseous products that were thought to be boron oxides (BO, B_2O_3) in early studies. However, a drastic increase of the purification rate is observed when hydrogen is present in the gases together with oxygen (eventually in the form of water vapor). In all cases with hydrogen, boron is removed as a hydroxide, the best candidate being HBO, as shown by Alemany et al. [37].

---

* With the latter still in development to reduce its cost.

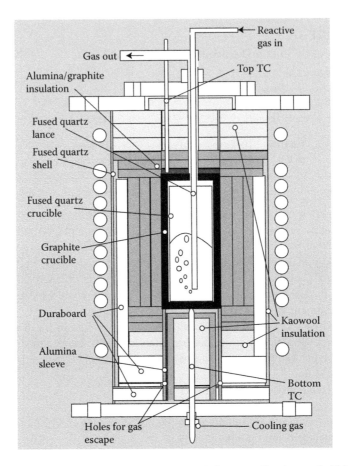

FIGURE 3.10 Schematic diagram of a bench-scale DS reactor allowing gas bubbling. (After C. P. Khattak, D. B. Joyce, and F. Schmid, 2001. Production of solar grade (SoG) silicon by refining liquid metallurgical grade (MG) silicon, *Final Report NREL/SR-520-30716*, Crystal Systems, Inc., April 19. Courtesy of C. P. Khattak.)

Thus, the boron removal reaction can be written[*]

$$B_{Si} + H_2O \rightleftharpoons HBO^v + \tfrac{1}{2}H_2 \tag{3.10}$$

This reaction is in competition with the oxidation of silicon, which should be suitably addressed at the production of SiO vapors and not of SiO_2, in order to avoid the formation of a slag that would block the reaction occurring at the surface between liquid silicon and the gas phase.

This condition would limit the oxidant contents in the gas phase [36,37] (and the oxidation rate), but this limit can be pushed back by increasing the temperature.

---

[*] As in earlier chapters, a chemical symbol in a reaction equation could bring a superscript indicating the thermodynamic state and a subscript indicating the nature of the phase in which it is dissolved.

(a)   (b)

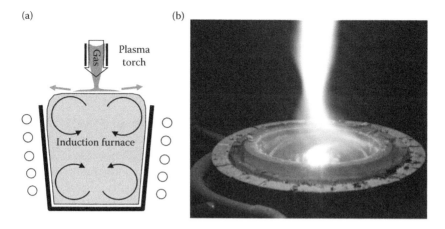

FIGURE 3.11 (a) Schematic drawing of a plasma reactor. (b) Details of a plasma reactor at laboratory scale where one sees a segmented cold crucible, the silica plates for thermal insulation, the graphite crucible with molten silicon, and the plasma.

Silicon oxidation models such as that of Ratto et al. [47] or its extension with hydrogen [48] are needed to understand and predict this limit.

3.2.3.1 Chemistry of Oxidative B Removal

In active oxidation conditions (direct gas–liquid contact), the analysis of the overall process in equilibrium conditions might be discussed by assuming, first, that the silicon oxidation reaction at the surface occurs with the formation of a volatile SiO species

$$\text{Si}^l + \text{H}_2\text{O} \rightleftharpoons \text{SiO} + \text{H}_2 \tag{3.11}$$

The competition between boron removal and silicon oxidation processes (reactions 3.10 and 3.11) might be controlled by chemical kinetics (depending on the active reaction path) or by some low transport processes of chemical species from the bulk to the surface or across the interface (see details in next section).

In the case of absence of reaction kinetics hindrances, the system is at its chemical equilibrium and the balance between the B removal and silicon oxidation is given by the equilibrium constant of the reaction

$$\text{SiO}^v + \text{B}_{\text{Si}} + \tfrac{1}{2}\text{H}_2 \rightleftharpoons \text{Si} + \text{HBO} \tag{3.12}$$

$$K = \frac{p_{\text{HBO}}}{p_{\text{SiO}} a_{\text{B(Si)}}} \, p_{\text{H}_2}^{-1/2} \tag{3.13}$$

where $a_{\text{B(Si)}}$ is the activity of B in liquid silicon relative to the equilibrium conditions of reaction 3.12.

For given conditions at the interface (partial pressure of SiO and H_2 pressure), the HBO pressure in the gas phase at equilibrium is thus proportional to the boron concentration in the liquid.

For a transport limited process with the interface at equilibrium, the boron removal rate is proportional to this equilibrium HBO partial pressure and thus to the boron concentration in the liquid, which gives a pseudo first-order reaction

$$r(\text{mol/sec}) = -\frac{d[\text{B}]}{dt} = k[\text{B}] \tag{3.14}$$

Such a first-order process was found experimentally in plasma [36,37] and in wet gas processes [44].

Furthermore, both the experimental values of the "enrichment factor," defined (and measured) in Reference 41 as the ratio of the boron concentration in the gas phase to the boron concentration in the liquid phase for a plasma process and of the rate constant k measured in Reference 44 for wet gas process, were found to satisfy conditions of thermodynamic equilibrium, showing that rate of the overall process is actually, not limited by reaction kinetics hindrances [48].

The thermodynamic equilibrium of reaction (3.12) depends on $K/p_{\text{H2}}^{1/2}$ and then on the formation enthalpy and entropy of HBO and SiO, and on the activity coefficient of boron in silicon [41]. Although only a limited amount of thermodynamic data is available for SiO [49], for HBO [43], and for the activity coefficient γ_{B} of B in Si (at 1723 K $\gamma = 0, 24$) [50], within the uncertainty of the known data, the calculated values relative to the equilibrium conditions of reaction 3.12 well correspond to the experimental values found for both a plasma [41] and for a gas blowing process [44].

For both process conditions, thermodynamics predicts, also, that the boron to silicon ratio should decrease when the temperature increases (see Figure 3.12), and that the purification rate increases as $p_{\text{H2}}^{1/2}$ when the partial pressure of hydrogen increases in the blown gases (see Figure 3.13).

Therefore, to increase the amount of B removal, the partial pressure of hydrogen in the blown gas should be increased (ideally, by suppressing the presence of argon or any neutral carrier gas), and the temperature should remain sufficiently low. However, lowering the temperature also lowers the maximum amount of oxygenated species that can be injected before encapsulating the liquid silicon with a layer of silica, as discussed earlier. Therefore, a compromise concerning temperature conditions has to be found for each process.

According to Tang et al. [43], a heterogeneous process occurring at the interface of silicon/gas, not reaction kinetics, controls the rate of the B-removal process.

The rate-limiting step has to be either the transport mechanisms of reactants to surface, or that of products from the surface to the reactor exhaust, according to the following paths:

- Boron transport in the liquid to the surface
- SiO or HBO transport from the surface to the bulk gas
- Oxidant transport from the bulk gas to the surface
- Overall flow of oxygen entering the reactor

Liquid phase transport has been studied numerically by Reference 51 and was found very fast, when compared to the purification rate in a laboratory plasma reactor, used with electromagnetic stirring of the bath.

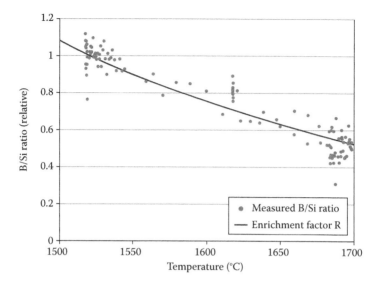

FIGURE 3.12   Variations of the B/Si ratio in liquid silicon and of the calculated enrichment factor with the temperature in a plasma process.

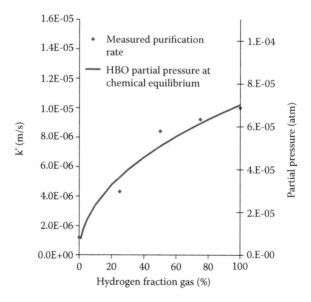

FIGURE 3.13   Purification rate in gas blowing experiments versus the hydrogen ratio. (Experimental data from E. F. Nostrand and M. Tangstad, 2012. *Metall. Mater. Trans. B*, 43B, 814–822.)

On the base of their experimental results,[*] Sortland and Tangstad [44,52] also found that liquid phase transfer is not rate limiting.

Other experiments with gas blowing inside the liquid [34], that promote a "vigorous stirring" of the melt, bring diffusion outcomes in the liquid phase to be negligible. Also in large size, industrial furnaces blowing inside the liquid will often promote some turbulent convection in the melt that can be enhanced with electromagnetic stirring. Thus, liquid phase transport is normally not rate limiting.

The transport of gaseous species across the gas-side boundary layer at the interface can be, instead, rate limiting, especially for blowing systems operating with a small relative velocity of the gas at the interface.

The mass transfer coefficient will be determined by the thickness of the dynamic boundary layer for surface blowing or by the size of bubbles for internal blowing, and will have roughly the same value for HBO, for the oxidizing species (H_2O or O), and for SiO because of their comparable diffusion coefficients in the bulk gas. This transport phenomenon is thought to be the rate-limiting step in a number of situations. Moreover, the transport of H_2O and silicon oxides becomes complicated if SiO is re-oxided into SiO_2 (solid particles) at some place in the boundary layer, which is certainly the case in most conventional process situations (except plasma processes because of their high gas temperature).

Also, the quantity of oxygen that is fed into the reactor provides a rate limitation: even for rapid transport in the melt and across the gas-side boundary layer, the total amount of SiO molecules produced is limited by the total amount of O atoms provided, or to half this amount if SiO is transformed into SiO_2 above the surface. With a system remaining in equilibrium conditions (Equation 3.13), the rate of boron removal can be deduced for a given boron concentration in the melt using Equation 3.14 and the appropriate rate constant. This gives a maximum value for the boron removal rate at a given oxygen (or water vapor) flow. Increasing it is only possible by increasing the enrichment factor by the total oxygen flow, which is limited by the formation of a silica layer at the surface.

3.2.4 ACID LEACHING

Cast MG-Si consists of a heterogeneous mixture of impurity-contaminated Si and of silicide (or carbide) phases of impurities present in supersaturated conditions. It is well known that silicon is insoluble in HCl mixtures, while at least silicides are soluble in it and might, therefore, be leached out.

HCl leaching can be used, therefore, in a purification process addressed at the removal of HCl-soluble impurities or compounds, but of the several leaching processes that have been tested, only that developed by Elkem [53] went to full success, as will be described in full detail in Chapter 4.

The main reason for this is that, even when MG-Si is finely ground, only the separate impurity phases, which are distributed randomly at the grain boundaries, might be dissolved by the acid mixture, while the silicon itself remains undissolved with the impurities at, or below, their equilibrium solubility.

---

[*] In these experiments, gas blowing at the surface was carried out using an induction heated graphite crucible.

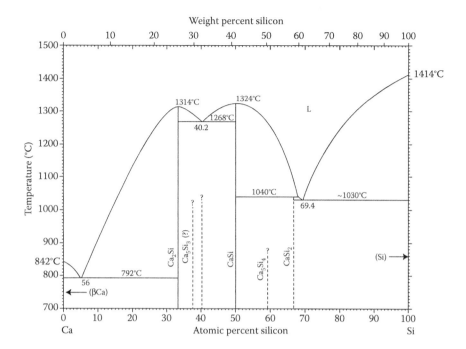

FIGURE 3.14 Phase diagram of the binary Ca–Si system. (From web image www.himikatus.ru.)

In the Elkem process [54], a liquid Ca–Si alloy is preliminarily prepared by adding a few percentages of Ca to MG-Si, which is subsequently cast in a suitable mold.

Different from the solidification of MG-Si, which occurs very close to the melting point of pure silicon, for the case of Ca–Si alloys (see phase diagram in Figure 3.14), the solidification initially proceeds with the segregation of silicon until the eutectic temperature is reached.

In this temperature range, the solidification occurs with the repartition of the impurities between the liquid and solid phase in pseudo-isothermal conditions, and the distribution ratio is given by the distribution coefficient k

$$k = C_s / C_l \tag{3.15}$$

where C_s is the concentration of the impurity in the solid and C_l is that in the liquid.

Table 3.2 reports literature data for the k values of most common impurities of MG-Si, from which one observes that for most metallic impurities $k \ll 1$, while it is 1 or close to 1 for C, O, B, and P.

In equilibrium conditions[*], the solidification process follows the classical Scheil's equation

$$C_s = kC_0(1 - f_s)^{k-1} \tag{3.16}$$

where C_0 is the initial impurity concentration and f is the fraction of liquid solidified.

---

[*] That is certainly not the practical case of the casting process of Ca-rich MG-Si.

TABLE 3.2

Segregation Coefficient of Selected Impurities in Silicon

Dopants[a]	Al	3×10^{-2}
	B	0.8
	P	0.35
Transition metals[a] (FZ)	Ag	5.0×10^{-7}
	Co	1.0×10^{-7}
	Cr	2.5×10^{-6}
	Cu	3.0×10^{-7}
	Fe	5.0×10^{-6}
	Mn	5.0×10^{-6}
	Mo	1.5×10^{-7}
	Ni	1.5×10^{-7}
	Ta	1.8×10^{-8}
	Ti	1.8×10^{-6}
	V	1.9×10^{-6}
	W	1.0×10^{-8}
	Zr	1.8×10^{-8}
Other[b]	C	0.05
	Mg	3.2×10^{-6}

Source: After G. L. Coletti, D. MacDonald, and D. Yang, 2012. Role of impurities in solar silicon, in *Advanced Silicon Materials for Photovoltaic Applications*, S. Pizzini (Ed.), Wiley, pp. 79–125.

[a] From FZ growth experiments.
[b] From CZ growth experiments.

As $k \ll 1$ for most of the metallic impurities the Scheil's equation, actually, simplifies to

$$C_s/C_l = k/(1 - f_s) \tag{3.17}$$

This means that in the early stage of the casting process, which implies the nucleation and growth of silicon crystallites, each single grain will present a concentration profile, with an increase of impurity concentration (for k values <1) from a pure core toward an impurity enriched surface.

At the eutectic temperature,[*] the process occurs isothermally, until the entire mass is solidified, with the simultaneous segregation of Si and a $CaSi_2$ phase, in the case

---

[*] Or peritectic temperature, depending on the multinarity of the impurity system.

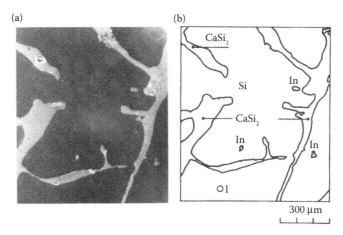

FIGURE 3.15   Transmission electron microscope (TEM) micrograph of a section of a leaching alloy (a) with details of the chemical composition (b). In means inclusions. (Adapted from A. Schei, 1985. High purity silicon production, in *Refining and Alloying of Liquid Aluminium and Ferro-Alloys*, T. A. Engh, S. Lyng, and H. A. Øye (Eds.), Aluminium-Verlag, Trondheim, Norway; A. Schei, 1986. Metallurgical production of high purity silicon, in *INFACON 86 Proceedings*, pp. 389–398.)

of a binary Ca–Si alloy, or of ternary $Ca_{2-x}Me_xSi_2$ or of multinary silicide phases in the case where silicon is contaminated by multiple impurities.

The microstructure of this heterogeneous solid is shown in Figure 3.15, where large islands of the $CaSi_2$ phase are distributed between the almost pure silicon grains. Several inclusions of different phases are, however, also present, which may present solubility problems.

It has been empirically demonstrated that the optimization of the composition of the Ca–Si alloy is essential to maximize the amount of metallic impurities segregated as mixed silicides in the calcium silicide phase.

When the alloy is reacted with a mixture of HCl and $FeCl_3$, the silicide phases are easily dissolved, leaving a rather pure silicon phase, where the concentration of a specific impurity should closely depend on the initial impurity concentration in the liquid and on their segregation coefficients.

Results of the effect of Ca alloying and leaching tests for selected impurities are reported in Table 3.3 that shows the potential of the calcium silicide route, well developed industrially by Elkem with the Silgrain process (see Chapter 4).

Different from other impurities, B and P could not be leached out with the use of Ca–Si alloys, due to the thermodynamic instability of boron silicide at temperatures above 1370°C and to the decomposition of phosphorous silicide at 1172°C due to the thermodynamic instability of boron silicide at temperatures above 1370°C and to the decomposition of phosphorous silicide at 1172°C.

3.2.5   SLAGGING PROCESSES

It has been already shown that B is an impurity that is very difficult to remove from a MG-Si feedstock using a liquid to solid extraction process. Molten salt mixtures,

TABLE 3.3

Experimental and Calculated Values of Leaching Experiments

	Fe (ppmw)	Al (ppmw)	Ca (ppmw)
Leaching alloy	3600	3700	29,000
Leached product	17	150	200
Calculated values	0.1	19	

Source: Data after A. Schei, 1986. Metallurgical production of high purity silicon, in *INFACON 86 Proceedings*, pp. 389–398.

consisting generally of silicates, or mixed silicates and fluorides, are instead shown to behave as effective media for the extraction of B from liquid silicon, where it is present at parts per million (ppm).

3.2.5.1 Thermodynamic Aspects[*]

The basic chemistry of the process, in the case of use of an oxidic slag, is the oxidation of the impurity at the liquid/liquid interface, with the formation of an oxide

$$Me_{Si} + O_{slag} \rightleftharpoons MeO_{slag} \tag{3.18}$$

which then dissolves in the liquid slag.

The thermodynamics of a slagging process are ruled by the activity of oxygen at the reaction interface. In turn, it depends on the activity of SiO_2 [55], according to the following equation:

$$O_{slag} + \tfrac{1}{2} Si_{Si} \rightleftharpoons \tfrac{1}{2} (SiO_2)_{slag} \tag{3.19}$$

and then on the slag composition (see Figure 3.16 for a binary SiO_2–CaO system).

In equilibrium conditions, the removal of B from Si at the reaction interface (int) is given by the following equation:

$$B_{Si} + \tfrac{3}{4} (SiO_2)_{int} \rightleftharpoons \tfrac{3}{4} Si_{Si} + (BO_{1.5})_{int} \tag{3.20}$$

and by the corresponding equilibrium constant

$$K_{int} = \frac{a_{Si_{Si}}^{3/4} \, a_{BO_{1.5}}}{a_{B_{Si}} a_{SiO_2}^{3/4}} \tag{3.21}$$

It should be noted that reaction (3.20) would be strongly shifted to the left when pure silica would be used as the oxidant, considering that $\Delta G_{SiO_2}^{o} \ll \Delta G_{BO_{1.5}}^{o}$. Therefore, basic fluxes are used to reduce the activity of SiO_2.

---

[*] The thermodynamic aspects of slagging processes are discussed in full detail also in Chapter 6.

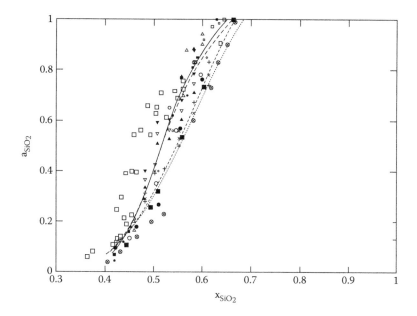

FIGURE 3.16 Activity of SiO_2 in binary $CaO-SiO_2$ systems. (Adapted from L. K. Jacobsen, 2013. Distribution of boron between silicon and $CaO-SiO_2$, $MgO-SiO_2$, $CaO-MgO-SiO_2$ and $CaO-Al_2O_3-SiO_2$ slags at 1600°C, PhD thesis, NTNU, ISBN 978-82-471-4791-7 [electronic version].)

For practical applications, it would be convenient to introduce mass concentrations in Equation 3.21, using $[\%B]_{Si}$ for the concentration by mass of B in the metal and $[\%B]_{slag}$ for the concentration of B by mass in the slag

$$K_{int} = \frac{a_{Si}^{3/4} a_{BO_{1.5}}}{a_{SiO_2}^{3/4} a_{B_{Si}}} = k^* \frac{[\%B]_{slag} \gamma_{BO_{1.5}}}{[\%B]_{Si} f_{B,Si}} \left(\frac{a_{Si}}{a_{SiO_2}} \right)^{3/4} \tag{3.22}$$

where $\gamma_{B_2O_3}$ and f_B are, respectively, the activity coefficients of $BO_{1.5}$ and B in the slag and metal phases. k^* is the coefficient for conversion from molar fraction to the mass concentration.

From the equilibrium constant of Equation 3.22, one could get the distribution coefficient L_B of B as the ratio between the B content in the slag and that in the metal

$$L_E = \frac{[\%B]_{slag}}{[\%B]_{Si}} = \frac{Kf_B}{\gamma_{BO_{1.5}} k^*} \left(\frac{a_{B_2O_3}}{a_{Si}} \right)^{3/4} \tag{3.23}$$

This distribution coefficient is a measure of the possibility for removal of B from Si.

Suzuki et al. [56] tested a number of different slag systems, which included the addition of CaF_2, for the removal of B under varying temperatures and in different atmospheres. They found that the distribution coefficient depends on the slag composition (see Figure 3.17) and that optimal distribution coefficients (L_B) exist for various fluxes (see Table 3.4) with a maximum value of 1.7 at 1500°C. They operated with an initial B content of 30–90 mass ppm and the experimental time varied between 1.8 and 10.8 ks. The experiments were conducted under a CO atmosphere at 1500°C. Similar experiments were performed by Tanahashi et al. [57] with comparable results (see Table 3.3).

The equilibrium distribution coefficients of B between the SiO_2–CaO–Al_2O_3 or SiO_2–CaO–Na_2O slags and liquid Si melt at 1823 K were also calculated using the new assessed thermochemical databank together with the FACT oxide thermodynamic database.[*]

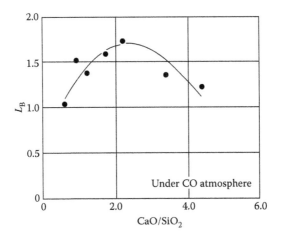

FIGURE 3.17 Experimental dependence of L_B on the CaO/SiO_2 ratio for the CaO–30%CaF_2–SiO_2 slag (solid line: calculated values setting unitary the activity of silicon). (After K. Suzuki et al., 1990. *J. Jpn. Inst. Met.*, 54, 168–172. Reproduced with permission of JIMM.)

TABLE 3.4

Values of L_B for Different Binary and Ternary Systems at 1500°C

System	CaO/SiO_2	L_B	References
CaO–10%MgO–SiO_2(–CaF_2)	~0.9	1.7	Suzuki et al. [56]
Ca–10%BaO–SiO_2(–CaF_2)	~1	1.7	Suzuki et al. [56]
CaO–30%CaF_2–SiO_2	~2.2	1.7	Suzuki et al. [56]
6%$NaO_{0.5}$–22%CaO72%SiO_2	~3.3	3.5	Tanahashi et al. [57]

[*] FactSage thermochemical software and databases, *Calphad*, 26, 189–228.

The SiO_2–CaO–Al_2O_3 slag seems to work better than the SiO_2–CaO–Na_2O slag. However, the calculated L_B values are approximately two times higher than the experimental values [4], showing that reaction kinetic barriers may play an important role in the refining processes, as will be discussed in the following section.

As an example, if we assume for L_B a value of 1.7, absence of B in the slag and 8 ppm B in the metal, the mass balance shows that 1.4 t of slag are required to reduce the B level from 8 to 1 ppm in 1 t of Si.

It is, in every case, apparent that the slag refining using silicate fluxes does not lead to a significant reduction of B content in MG-Si feedstock with a single run. Multiple runs are, therefore, needed on demand. Better results could be, however, obtained, using slags that favor the formation of an insoluble compound of B, as suggested by L. Pelosini et al. [58].

3.2.5.2 Kinetic Aspects

When a dissolved element is subjected to a slag treatment aimed at its removal, as illustrated in the former section, it goes through the following five steps [59] (see also Figure 3.18):

1. The impurity element X must be transferred from the bulk metallic phase to the metal boundary layer, $[X]_b \rightarrow [X]_\delta$
2. The impurity element must diffuse through the metal boundary layer, $[X]_\delta \rightarrow [X]_i$
3. The metal is oxidized at the interphase between metal and slag, $[X]_i \rightarrow (X)_i$
4. The impurity element diffuses through the slag boundary layer, $(X)_i \rightarrow (X)_\delta$
5. The impurity element is transferred from the slag boundary layer to slag bulk phase, $(X)_\delta \rightarrow (X)_b$

The rate of steps 1 and 5 depends on the proper stirring and mixing in the metal and slag phases. Although there is a serious difficulty of mixing the impurity species in the slag phase that is often very viscous, it is currently assumed the slag phase is "completely mixed."

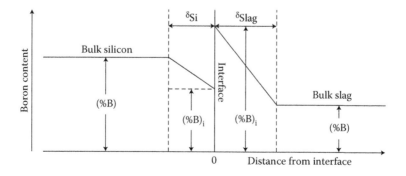

FIGURE 3.18 Concentration profile of B in the silicon and the slag, where boron is transferred from silicon to the slag.

Stirring is often carried out by gas bubbling or using mechanical devices to increase the mass transfer in the bulk phases. Hence, the thermophysical properties of the slag, such as its density and viscosity, are important to be well known to optimize the process. A high viscosity would be responsible for a low mass transfer of the impurity element.

Steps 2 and 4 depend on the mass transfer coefficient in the metal, k, and in the slag, k_s, respectively. With the assumptions that steps 1, 5, and 3 are much faster than steps 2 and 4, the final concentration depends only on the transfer coefficients of the impurity across the boundary layers.

The solution of the problem may be obtained by considering, first, that the accumulation rate of the impurity in the slag corresponds to its removal rate through the melt boundary layer

$$-\frac{d[\%X]_e}{dt} = k_s \rho A_s \left([\%X]_s - [\%X]_e \right) \tag{3.24}$$

where k_s is the rate constant, ρ is a resistance factor to the mass transfer, and $[\%X]_e$ is the hypothetical concentration of X in the metal in equilibrium with the actual concentration in the slag $[\%X]_s$ so that $([\%X]_s - [\%X]_e)$ is the driving force, and

$$[\%X]_e = \frac{\gamma_x[\%X]_s}{Kf_x} \tag{3.25}$$

To integrate Equation 3.24 over time to get the final concentration of X in the silicon melt, $[\%X]_e$ must be replaced by a function of $[\%X]_s$. $[\%X]_e$ may be obtained in terms of $[\%X]_s$ if we note that whatever leaves the melt enters the slag

$$M_e([\%X]_e^o - [\%X]_s) = M_s[\%X]_s \tag{3.26}$$

where M_e is the mass of silicon, $[\%X]_e^o$ is the initial concentration of X in silicon, and M_s is the mass of the slag. Here, we assume that the slag originally did not contain any component X. Thus, the above equation together with Equation 3.25 gives

$$[\%X]_e = \frac{\gamma_x}{f_x K} \frac{M_e}{M_s}([\%X]_e^o - [\%X]_s) \tag{3.27}$$

The driving force from Equation 3.27 becomes

$$[\%X]_s - [\%X]_e = [\%X]_s \left(1 + \frac{\gamma_x M_e}{Kf_x M_s} \right) - \frac{\gamma_x M_e}{Kf_x M_s}[\%X]_e^o \tag{3.28}$$

The lowest value of $[\%X]_e$ attainable, that is, the lowest content of the impurity X in liquid silicon is when the driving force given by Equation 3.28 becomes zero. Then

$$[\%X]_s = [\%X]_s^\infty \frac{\gamma_x M_e}{(Kf_x M_s + M_e)} [\%X]_e^o \qquad (3.29)$$

where $[\%X]_s^\infty$ is the value of $[\%X]_s$ when equilibrium between slag and melt is finally reached (at time $t \to \infty$).

This gives, on integration of Equation 3.24 the results schematically reported in Figure 3.20, assuming that k_t, y_x, and f_x are independent of time. In Figure 3.19, we see that $[\%X]$ drops exponentially down to the value $[\%X]_\infty$, corresponding to equilibrium conditions of X in the melt and in the slag.

Here, we have assumed that resistance is in the melt boundary layer. The same procedure could be repeated for the case of a resistance in the slag boundary layer with similar results.

3.2.6 IMPURITY TRANSFER BY LIQUID–SOLID SEGREGATION: ALUMINUM SMELTING

Al is a formidable solvent of silicon (see Figure 3.20a), while being, instead, almost insoluble in solid silicon (see Figure 3.20b), with a solubility maximum of 0.02% around 1180°C (2.9×10^{19} at cm^{-3} at 1450 K), above which a retrograde solubility behavior does occur [60].

The most interesting feature of this system is its potential to allow an efficient repartition of metallic impurities between Al–Si melts and solid silicon [61–64], due to the very small values of their segregation coefficients (see Table 3.5) that makes Al–Si melts a convenient medium for low temperature MG-Si purification [1, pp. 21–78].

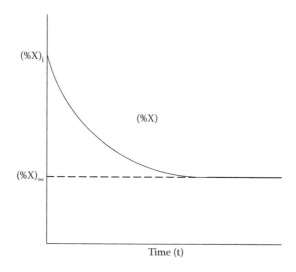

FIGURE 3.19   Removal of an impurity X from metal to slag.

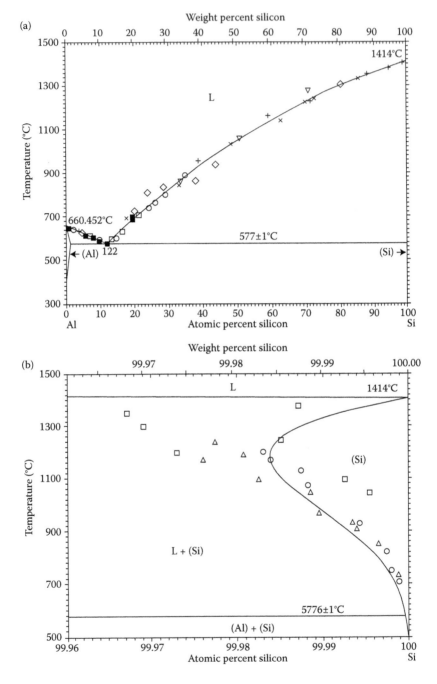

FIGURE 3.20 (a) Phase diagram of the Al–Si system, (b) the solubility of Al in solid silicon. (With kind permission from Springer Science+Business Media: *Bull. Alloy Phase Diagr.*, The Al–Si (aluminum–silicon) system, 5(1), 1984, 74–84, J. L. Murray and A. J. McAlister, Springer Copyright licence nr. 3720130976003.)

TABLE 3.5

Temperature Dependence of the Segregation Coefficients of Metallic Impurities in Al–Si Alloys

Element	1073 K	1273	1473	m.p. of Si
Fe	1.7×10^{-11}	5.9×10^{-9}	3.0×10^{-7}	6.0×10^{-6}
Ti	3.8×10^{-9}	1.6×10^{-7}	9.6×10^{-7}	2.0×10^{-6}
Cr	4.9×10^{-10}	2.5×10^{-8}	2.5×10^{-7}	1.1×10^{-5}
Ni	1.3×10^{-9}	1.6×10^{-7}	4.5×10^{-6}	1.3×10^{-4}
Cu	9.2×10^{-8}	1.2×10^{-7}	2.1×10^{-6}	1.0×10^{-5}
B	7.6×10^{-2}	2.2×10^{-1}		
P	4.0×10^{-2}	8.5×10^{-2}		

Source: Data from T. Yoshikawa and K. Morita, 2005. *ISIJ Int.*, 45, 967–971; T. Yoshikawa and K. Morita, 2009. *J. Cryst. Growth*, 311, 776–779.

It is, eventually, interesting to observe (see Figure 3.21) that B and P segregation works better in Al–Si melts than in liquid silicon, and increases while decreasing the temperature, also making possible reasonable B and P removal, otherwise impracticable.

The main challenge of the Al-smelting process is, however, to bring the residual concentration of Al in the purified silicon feedstock below to about 0.2 ppma. In

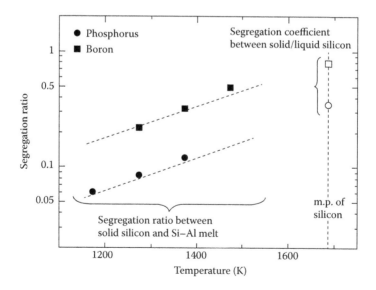

FIGURE 3.21 Segregation coefficients of B and P in a Al–Si melt. (Reprinted from *J. Crys. Growth*, 311, T. Yoshikawa and K. Morita, Refining of silicon during its solidification from a Si–Al melt, 776–779, Copyright 2009, with permission from Elsevier.)

fact, Al is a dopant, but also a recombining impurity [65], that behaves as a minority carrier killer at concentrations above 10^{16} at cm^{-3} [1, pp. 53–54] [66]. On the other side, a process of Al purification using DS suffers an inappropriately high segregation coefficient of Al at the melting point of Si (3×10^{-2}) [67, pp. 79–125]. Therefore, Al removal should be carried out using an ad hoc designed crystallization process, that involves the simultaneous occurring of a slagging step at the top surface of the silicon melt charge [68], capable to segregate Al efficiently in the slag.

A practical way to use Al smelting for the production of purified silicon is to work with MG-Si–Al melts, solidified at temperatures lower than that corresponding to the solubility maximum, typically 1273 K or lower [62] to keep the Al concentration in the segregated Si as low as possible and to favor the B and P removal. In these conditions, the excess silicon segregates as a separate phase of platelet (flakes) Si crystals, uniformly distributed into the solidified structure [62]. A purified silicon sponge could then be recovered by acid etching at room temperature of the entire solidified mass,* or the silicon flakes could be separated by agglomeration on the bottom of the furnace when the alloy is solidified in an induction field [62]. Depending on the initial sample composition, the experimental Fe content ranges between 13 and 47 ppmw, that of Ti between 2.7 and 7.7, of B between 0.81 and 0.99, of P between 0.77 and 1.2, and that of Al between 453 and 599 ppmw.

A different practice that brings much better results, is followed by Silicor [68] in its production process brought recently to the preindustrial stage. Here, in a first step, Si is crystallized from a hypereutectic Al–Si melt by slow cooling, that is interrupted just before the eutectic point is reached. The excess silicon crystallizes with a residual content of metallic and dopant impurities ruled by the segregation equilibrium

$$\mu_{Al,l}^{Me} = \mu_{Si,s}^{Me} \tag{3.30}$$

where μ is the chemical potential of the metallic impurity Me. The remaining melt after this step has around 83 wt% Al, 16 wt% Si, and 1 wt% of other impurities and is poured away leaving behind a silicon sponge which is carefully acid leached to remove the residual Al–Si eutectic covering the silicon surface.

The final process is a single crystallization growth to reconduct the feedstock to SoG purity.†

According to M. Heuer et al. [69], the material obtained after a solidification growth presents an Al content around 0.1 ppmw, and could be used with success as a solar silicon feedstock (see details in Section 3.3.2).

3.2.7   IMPURITY TRANSFER BY LIQUID–SOLID SEGREGATION (DS GROWTH AND CZOCHRALSKI PULLING)

The final purification of a solar feedstock arising from MG-Si is currently carried out by DS, although in the early projects Czochralski (CZ) growth was used, as DS

---

* A very expensive process.
† Section 3.3.2 reports details of the preindustrial and industrial processes developed so far using Al smelting.

furnaces addressed at the growth of multicrystalline (mc) silicon ingots were not yet available.

Among the advantages of DS growth for purification purposes, the large ingot weight (up to 500 kg) and the overall process costs are the main issues, together with a larger tolerance in the initial total impurity content.

In both the DS and CZ case,[*] the process is based on the repartition of impurities between the liquid and solid phase. In equilibrium conditions, the distribution is ruled by distribution (or segregation coefficients) k, which, as already shown in Section 3.2.4, is given by the ratio

$$k = C_s/C_l \qquad (3.31)$$

with C_s as the concentration of the impurity in the solid phase and C_l as the concentration of the impurity in the liquid phase.

It is easy to demonstrate [67] that the effective values of the segregation coefficients k_{eff} depend on the type of growth process used, and are a function of the growth rate v and of the thickness δ of the diffusion layer which sets up at the interface between the liquid and the solid phase

$$k_{eff} = \frac{1}{1 + ((1/k_{eq}) - 1)\exp(-(v\delta/D))} \qquad (3.32)$$

as well as of the diffusion coefficient D of the specific impurity involved in the segregation process.

Since for the majority of metallic impurities $k \ll 1$ (see Table 3.2), the impurities segregate mostly in the liquid phase, and liquid–solid segregation is thus an effective, and systematically applied, purification process [1,67].

For the majority of metallic elements, assuming complete mixing in the liquid phase and no practical diffusion in the solid phase, the solidification follows the Scheil's equation

$$C_s = kC_0(1 - f_s)^{k-1} \qquad (3.33)$$

where C_0 is the initial concentration in the liquid and f_s is the fraction of liquid solidified.

It is, therefore, apparent that an impurity profile along the ingot is expected, with an increase of the impurity concentration for $k_{eff} < 1$ and a decrease of the impurity concentration for $k_{eff} > 1$.

As the equilibrium and effective segregation coefficients for B and P are close to 1, and that of oxygen is equal to 1, for these impurities the solidification process is not a purification tool.

---

[*] The reader interested in major details concerning the DS and CZ growth of semiconductor silicon may refer to Reference 12.

3.2.8   ELECTROCHEMICAL PURIFICATION

3.2.8.1   Electrochemical Background

At least theoretically, electrochemistry should provide a straightforward mean to get pure silicon by electrolysis of a melt in which high purity quartz has been dissolved or by anodic dissolution of MG-Si and Si electrodeposition.

This approach, however, suffers from a number of challenges, which, at the time of writing this chapter, are still unresolved. As an example, the direct use of solid MG-Si anodes is generally precluded, due to the onset of electrode passivation phenomena associated with silicon oxidation[*] [70]. Moreover, electrodeposition of liquid silicon is a high temperature operation in fluoride melts,[†] associated to serious construction material problems. Eventually, solid silicon electrodeposition occurs under severe overvoltages, unless working at very low current densities, which leads to microcrystalline deposits that easily detach from the electrode, independently of its nature, during the electrolytic process.

Last, but not least, the electrochemical properties of elements in fluoride melts are only partially known,[‡] although a series of electrochemical potentials of elements in LiF–NaF melts at 1173 K (see Table 3.6) has been proposed by S. Fabre et al. [72] as an integration of the data reported by Olsen and Rolseth [73], see Table 3.7.

From these data, one could remark that Ca, Mg, and Al are more electronegative than silicon and should preferentially dissolve from a Si alloy containing these elements, leading to a potential purification process, while the B purification would be precluded, as its deposition potential

$$E = E^\circ - \frac{RT}{3F} \ln[B^{3+}] \tag{3.34}$$

at its normal concentration in MG-Si (≥ 50 ppmw), turns out to be very close to that of silicon (see Table 3.7).

It should be also noted that a partial dissolution of the more electronegative elements present in a Si-alloy would occur also without the application of current, with

---

TABLE 3.6

Electrochemical Potentials of Metals versus a Pt Pseudo-Reference Electrode in LiF–NaF Melts at 1173 K

	Na	Cr	Fe	Ni	Mo	W	Ag	Au
E/Pt (V)	−1.40	−0.22	+0.04	+0.46	+0.65	+0.90	+1.10	+2.15

Source: Data from S. Fabre et al., 2013. *J. Nucl. Mater.*, 441, 583–591.

---

[*] In mixed halide–oxide melts, the anodic dissolution of silicon could be associated to oxygen evolution.
[†] Which are good solvents for silica.
[‡] Electrochemical measurements in fluoride melts using oxygen and hydrogen reference electrodes were carried out by one of the authors of this chapter [71].

TABLE 3.7
Electrochemical Series of Elements in Fluoride Melts

	P/PF$_5$	Fe/FeF$_2$	Fe/FeF$_3$	P/PF$_3$	B/BF$_3$	Si/SiF$_4$	Al/AlF$_3$	Mg/MgF$_2$	Ca/CaF$_2$
E°(V)	2.363	2.529	2.621	2.704	3.538	3.551	3.698	4.176	4.899

Source: Data according to E. Olsen and S. Rolseth, 2010. *Metall. Mater. Trans.*, 41B, 295–302.
Note: The reference is a fluorine electrode.

an electroless process driven by the difference of the electrochemical potentials of silicon and metallic impurities.

This system behaves, in fact, in the presence of silicon ions Si^{4+} in the electrolyte, like a galvanic cell in short circuit, with Si working as the cathode and the metallic impurities at the surface of the sample as the anode.

The overall rate of this currentless process would, obviously, depend on the rate of the slow partial process, here the cathodic processes, ruled by the reduction of silicon ions Si^{4+} in the fused electrolyte*

$$Si^{4+} + 4e \rightleftharpoons Si \qquad (3.35)$$

taking the oxidation of Ca (as well as Mg and Al) impurities in the alloy as the fast anodic process

$$2Ca \rightleftharpoons 2Ca^{2+} + 4e \qquad (3.36)$$

The electroless process would play, however, only a minor role, unless reducible substances would deliberately be added to the melt, as would be the case of melts with hexafluoride $[SiF_6]^{2-}$ ions.

In the case of MG-Si–Cu alloys, which will be discussed later in this section, electroless conditions could work even better, as Cu works as the cathode and impure Si will work as the anode and would spontaneously dissolve together with its impurities, provided reducible ions are available in the electrolyte.

3.2.8.2 Electrochemical Processes

The first attempt to prepare Si using an electrochemical process was by Ullik who electrodeposited silicon from a molten bath of K_2SiF_6–KF in the temperature range 600–900°C [74]. Later authors suggested that cryolite (Na$_3$ AlF$_6$) melts containing silica could be suitable media for the electrodeposition of silicon and of Al–Si alloys, with a process that could bring to the production of silicon and of its alloys in a single carbon-free step, resembling the Hall–Heroult process of Al production.

---

* We will show in the next section that the reduction of Si^{4+} to silicon occurs with low overvoltages, but we assume here that the anodic process would be in any case faster.

The solubility of silica in pure cryolite melts is, however, low (<5% at 1010°C), but it could be increased in $Na_3 AlF_6–Al_2O_3$ mixtures. The physicochemical properties of $Na_3 AlF_6–Al_2O_3–SiO_2$ mixtures were investigated, among others, by F. Grjotheim et al. [74], who determined also the experimental value for the decomposition voltage of SiO_2 in a cryolite melt added with 5% of silica (1.10 ± 0.1 V on a graphite electrode at 1273 K). This value is much lower than the calculated standard potential (1.675 V at 1300 K), on the hypothesis that the anodic reaction at a graphite electrode is the carbon oxidation to CO_2.

A pilot plant addressed at the electrolytic production of silicon from a SiO_2-cryolite melt was operated in the 1960s by R. Monnier [75]. A schematic drawing of the furnace used is reported in Figure 3.22, where one can remark that the crucible walls are covered by a solid layer of the electrolyte, which makes corrosion damages negligible, while heating was provided by the electrolytic current across carbon electrodes. Current densities up to 0.8 A/cm^2 were applied, comparable to that used in the conventional Al process, but the effective current density under which Si deposition occurs is much less.

Under these conditions, the product consisted of small silicon crystals, few millimeters in size, which only partially adhered at the carbon electrodes.

The project failed to reach industrialization, probably as a consequence of the low deposition rates, typical of solid silicon deposition (see later for details) [76] and the difficulty in collecting the silicon powder, that once melted and zone refined, was, however, of semiconductor grade according to the authors.

Since in $Na_3AlF_6–Al_2O_3–SiO_2$ mixtures, Al is co-deposited with Si [74], the feasibility of a pure silicon deposition using cryolite melts is excluded, and then this process is used for SoG production.*

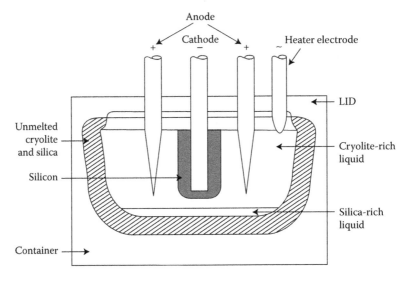

FIGURE 3.22 Monnier's electrolytic cell. (With kind permission from Springer Science+ Business Media: *J. Appl. Electrochem.*, Electrolytic production of silicon, 18, 1988, 15–22, D. Elwell and G. M. Rao, and Licence nr. 3720281208059.)

---

* The role of Al as a recombining impurity is discussed in Section 3.2.6.

A better, Al-free electrolyte, which could be potentially used for the electrolytic purification of Si,[*] is a mixture of K_2SiF_6 in a fluoride melt,[†] because $[SiF_6]^{2-}$ is a soluble complex of Si that dissociates in fluoride (or mixed chloride–fluoride melts)

$$[SiF_6]^{2-} \rightleftharpoons Si^{4+} + 6F^- \qquad (3.37)$$

leading to silicon ions that could be electroreduced [77] without significant overvoltage, provided the current density could be kept quite low. In this and in the following case, K_2SiF_6 is actually the silicon feedstock, and the overall process would imply its preparation from a quartz supply.

As shown earlier by Cohen and Huggins [78], the key point is, however, that working at small current densities to bring the overvoltage small, leads to very small practical deposition rates. They used a KF–LiF melt, added with 5% molar K_2SiF_6 at 750°C, with electrodes consisting of single crystal Si. The deposition rate was found of the order of 1 μm/h, too low to be interesting for industrial applications, in agreement with the DeMattei and Feigelson results [79] that show that the critical current for solid silicon electrodeposition is around 40 mA/cm², while current densities of 1 A/cm² are routinely employed in the case of liquid Al.

Additional details on the silicon electrodeposition in fluoride melts were given by G. M. Rao et al. [80], who carried out the electrodeposition of silicon on a silver electrode from a ternary LiF–NaF–KF eutectic at 750°C, with K_2SiF_6 additions in the range 0.5–6 mol%.

At the highest K_2SiF_6 additions, which are essential to get successful Si deposition, relatively thick (3 mm) and inclusions-free deposits were obtained, with a current efficiency ranging from 35% to 70%, with a projected power to silicon mass value of 16 kWh/kg at 750°C. The fact that the deposits were dense and coherent was used by the authors as the proof that the silicon deposition does not occur with the primary formation of alkali metal, which then would react with Si^{4+} ions

$$4K + Si^{4+} \rightleftharpoons Si + 4K^+ \qquad (3.38)$$

to give just Si as the final product, but is associated to the *direct* reduction of the complex species $[SiF_6]^{2-}$. In the best experimental conditions, the total impurity concentration was 10 ppmw, with B at 2 ppmw. The same authors carried out electrodeposition experiments on graphite electrodes [81] using the LiF–NaF–KF eutectic mixed with 8%–14% molar K_2SiF_6. Dense, coherent deposits were obtained, with a total content of metallic impurities (Cu, Fe, and Ni) around 50 ppmw.

Apparently, none of these attempts seems to be appropriate for the industrial development of a silicon refining process, being limited by inherent difficulties in managing the electrodeposition of solid silicon with proper purity and at sufficiently high rates.

The same practical limits were observed when using quartz electrodes and alkali- or alkali-earth halide melt as the electrolytes, which are also good solvent for silica

---

[*] Using an impure silicon anode.
[†] The same originally used by Ullik.

and for K_2SiF_6. $CaCl_2$ and fluoride melts offer this opportunity, with the advantage of fluoride melts to dissolve larger amounts of K_2SiF_6.

The few efforts carried out very recently with the direct use of quartz electrodes (see Figure 3.23) and a $CaCl_2$ electrolyte [82–84], although of some basic interest, remain of difficult practical application. The direct electroreduction of silica, in fact, does occur using pure silicon electrodes in contact with the quartz sample at a test temperature of 1123 K, operating the system in dry argon and using a quartz crucible as the melt container. The electroreduction products are columnar microcrystals of pure silicon.

Instead, in fluoride, chloride, mixed fluoride/chloride, or mixed halide/oxide melts silica should be used as the feedstock and directly reacted as such, in the presence of K_2SiF_6, which, as shown before, ensures the presence of reducible Si^{4+} ions and promotes as well the solubility of the silica, via an intermediate oxyfluoride complex.

Depending on the melt composition, the electrolysis could be carried out at a temperature above or below the melting temperature of silicon.

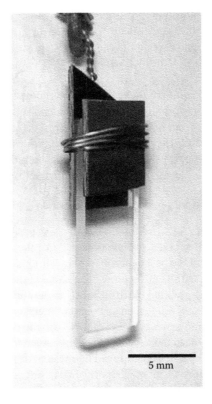

5 mm

FIGURE 3.23   The 5 mm wide, quartz electrode used in the experiments carried out by Yasuda et al. (Reprinted from *Electrochimica Acta*, 53, K. Yasuda et al., Direct electrolytic reduction of solid SiO_2 in molten $CaCl_2$ for the production of solar grade silicon, 106–110, Copyright 2007, with permission of Elsevier.)

The high temperature route was followed by Rao et al., with a $BaO-BaF_2-SiO_2$ melt as the electrolyte [81], at temperatures slightly below (1393°C) and above 1465°C, the melting point of silicon, using graphite electrodes.

As the density of silicon is close to that of the melt, in both cases the electrode-posited silicon remains adherent at the electrodes. The low Faradaic yield (40% in the best case) was explained as due to loss of SiO by sublimation and to the partial electronic conductivity of the melt, due the presence of a suspension of carbon particles detached from the electrode.

Chloride melts added with CaO are claimed to be suitable solvents for SiO_2 [70], but when tested for B and P purification, the success was modest [85].

Eventually, also carbonate melts are better solvents for oxides than chlorides, but their decomposition potentials are lower than that of silica, thus limiting the range of applicability of carbonates to the carbides electrosynthesis [86] or, by a suitable modification of the melt composition, to the electrodeposition of transition metal-borides, -carbides, and -silicides in halide melts [87,88].

Low Faradaic yield, powder form of the silicon produced, and modest quality of the electrodeposited silicon are, therefore, serious obstacles to the development of industrial processes based on conventional electrochemical processes using solid anodes and cathodes.

3.2.8.3 Electrorefining and Electrowinning Processes

Two different processes might be foreseen to produce pure silicon with an electrochemical process.

The first would be the preferential anodic dissolution of the more electronegative impurities from a MG-Si anode, which could then be melted and directionally solidified to remove the remaining metallic impurities. The preferential anodic dissolution is typically an electrowinning process.

The second would be the direct production of a pure silicon material by the anodic dissolution of a MG-Si source in a suitable electrolyte and the simultaneous deposition of silicon on an appropriate cathodic material.

It is clear that if the anodic dissolution of MG-Si occurs together with the contaminating impurities, the cathodic deposition would occur from an impurity-contaminated melt. One should, therefore, expect that the electrodeposited silicon should also be impurity contaminated.

Having this challenge in mind, several efforts were carried out to minimize or even suppress the electrolyte contamination, by using suitable anodic alloys or proper cell configurations.

The most interesting advances in this field were obtained, with

- Electrorefining processes using liquid anodes and cathodes[*] [89]
- Electrowinning processes using anodes solidified from a hypereutectic solution of copper and MG-Si, which behaves as a two-phase mixture of silicon embedded in a Cu_3Si matrix [90] or

---

[*] As a late realization of the suggestions of Monnier [75].

- Using additional electrocatalytic procedures, capable of inactivating impurities otherwise impossible to remove [73,91], with the application of the three-layer electrochemical refining

As said before, electrowinning processes are based on the preferential anodic dissolution of the less noble components of an alloy working as an anode in a proper electrolyte. In the case of silicon, metallic impurities more electronegative than silicon will preferentially dissolve.

After electrowinning, the purified silicon could be subjected to a DS process to remove the noble impurities dissolved in it. Electrowinning processes could be, however, associated to a silicon electrodeposition process from the same electrolyte.

The electrowinning process has been successfully applied for the preparation of 6N aluminum, but the attempts carried out at the Norwegian School of Science and Technology (NTU) and SINTEF in Norway [92–94] for the purification of MG-Si in $KF–LiF–K_2SiF_6$ melts resulted only in moderate success.

Cathodic silicon deposits are, in fact, heavily contaminated with C, O, and F, and a subsequent treatment in a CaF_2 melt was needed to get a Si powder still heavily contaminated with Fe (>1000 ppmw). The current efficiency is greater than 49%, with an energy consumption of 17 kW h/kg.

The process developed by Olson and Carleton [90] using anodes solidified from a hypereutectic solution of copper and MG-Si, which behaves as a two-phase mixture of silicon embedded in a Cu_3Si matrix, presents, instead, a considerable interest. In fact, the process works with a high Faradaic efficiency (up to 99%) at 750°C and yields a 99.999% pure electrodeposited Si, with the almost total removal of B and P (see Table 3.8).

The key issue of this process is that not only the more electropositive impurities that remain at the anode, as theoretically expected, but also the more electronegative impurities (Ca, Al, Cr, Ti, V), blocked by a diffusional trapping process, due to the

---

TABLE 3.8

Impurity Content in MG-Si and Electrorefined Si

Metal Impurity	Impurity Content in MG-Si (ppma)	Impurity Content in Electrodeposited Si (ppma)
Al	3400	1.0
B	17	0.7
Ca	290	<0.07
Mg	85	0.9
Cr	40	<0.2
Fe	>2500	0.1
Ti	290	<0.1
V	250	<0.05
P	7	<0.03

Source: Data from J. M. Olson and K. L. Carleton, 1981. *J. Electrochem. Soc.*, 128, 2698–2699.

---

presence of the Cu_3Si matrix which works as a diffusion filter. Therefore, impurities do not accumulate in the electrolyte and the simultaneous Si-electrodeposition at a suitable cathode occurs from an almost uncontaminated electrolyte.

The process developed by Olsen [73,91] is based on the use of a three-layer cell with a liquid Cu–Si (17% Cu) alloy as the anode, a (liquid) silicon as the cathode and a fluoride electrolyte consisting of a mixture of CaF_2 and BaF_2 with the addition of SiF_6^{2-} as an Si-carrying complex ion. The electrolyte composition is selected to get a density intermediate between the Cu–Si alloy, working as the anode and pure liquid Si, working as the cathode. At temperatures above the melting temperature of Si,[*] therefore, the Cu–Si alloy stays on the bottom and the liquid Si just on top.

As already mentioned before, all impurities less noble than Si (Ca, Ba, Mg, Al) contaminating a Si electrode in equilibrium with a fluoride electrolyte should dissolve while the more noble remain in the anode. B has an electrochemical potential[†] that is very close to that of Si ($E^\circ_{Si} = -3.551$ V, $E^\circ_B = -3.538$ V) but the dissolution voltage is slightly more negative, depending on its concentration. Some results reported in Table 3.9 show that a preferential Al and Ca dissolution occurs, which is instead negligible for iron and limited for boron.

The impurity content of Si deposited at the cathode, from an electrolyte strongly contaminated with Ca and Al, ranges for Cu around 200 ppmw,[‡] for Al around 3000 ppmw, for B between 3 and 10 ppmw, for P between 1 and 2 ppmw, for Fe <10 ppmw, and for Ti between 10 and 50 ppmw. The overall impurity contamination is, therefore, far from a SoG material.

In a further work, Olsen et al. [91] showed that it is possible to inactivate, at least partially, B during the anodic process with the electrocatalyzed formation of a titanium boride, which is insoluble in the Cu–Si alloy. It was shown, in fact, that at low polarizations, the formation of intermetallic borides occurs at the electrolyte/Cu–Si alloy interface.

TABLE 3.9

Impurity Concentration at the Anode after Electrowinning

Impurity	Initial MG-Si Concentration (ppmw)	Concentration at the Anode (ppmw)
Al	1200–4000	30
Ca	590	47
B	20–45	12
P	27–30	31
Fe	1600–3000	1300

Source: Data from E. Olsen and S. Rolseth, 2010. *Metall. Mater. Trans.*, 41B, 295–302.

---

[*] A graphite crucible lined with sintered Si_3N_4 was used as the container.
[†] With reference to a fluorine electrode.
[‡] The heavy Cu contamination is felt to be due to Cu vapors sublimating from the bottom electrode and diffusing toward the anode.

It is, however, apparent that electrowinning of MG-Si presents a number of challenges, mostly due to the contamination of the electrolyte arising by impurities anodically dissolved, unless an impurity diffusion barrier is used at the anode. In general, the materials so far obtained are far away from SoG properties and would neither be a suitable feedstock for a subsequent DS process.

3.3   PROCESSES

3.3.1   FERROATLANTICA: EFFORTS IN THE FERROATLANTICA GROUP (FERROSOLAR AND PHOTOSIL)

Ferroatlantica is the world's largest silicon metal producer, and one of the largest ferroalloy producers, with factories in Brazil, China, France, Venezuela, Canada, South Africa, and Spain. The Ferroatlantica's group I + D Company is its R&D Company, with several fields of interest, including solar silicon development as one of the main targets. The activities in this field are carried out within the FerroSolar and Photosil projects, the last carried out in France with the cooperation of FerroPem, Apollon Solar, CNRS, and CEA (Commissariat a l'Energie Atomique).

3.3.1.1   FerroSolar Silicon Project

Since 2000, the Ferroatlantica group has been working on a project upgrading MG-Si to SoG-Si using metallurgical methods. From the beginning, the FerroSolar project was integrated in the silicon metal (MG-Si) production facility of Ferroatlantica S.A. in Sabón, Spain, with the aim of using the technologies and knowledge developed not only to purify silicon to SoG but also to improve the production and the quality of the silicon metal.

Different from projects discussed in Chapter 2, aimed at the production of a high quality MG-Si from pure raw materials,[*] the FerroSolar project intends to integrate a conventional, but optimized, MG-Si production with a sequence of conventional purification steps.

A pilot plant started with standard induction furnaces in a size to process 4 t of steel which corresponds to approximately 1 t of silicon [94] and meanwhile was equipped with advanced furnaces to carry out vacuum treatments and directional solidification (DS) processes [16,95]. The total capacity of this facility is 400 t of SoG-Si per year and in 2014, about 150 t had been processed.[†]

The purification process starts with MG-Si of a better than standard quality, that is produced in special production campaigns. This special MG-Si is refined by a slagging process in induction furnaces. The slagging process is repeated as many times as necessary to obtain the target dopant concentrations for this step. In the first generation CS Silicon™ (GEN1) purification process, the silicon proceeds directly to the DS process after slagging. This step can be also carried out in an electromagnetic casting furnace which allows a continuous casting process and removes residual

---

[*] That was the main reason for a lack of success.
[†] J. Bullon, private communication, 2014.

carbon (SiC particles) as well as residual slags. Then, the resulting ingot is cut into bricks which undergo a final inspection and cleaning treatment [96].

In the second generation purification process (GEN2), an additional vacuum purification step has been introduced between slagging and solidification to reduce the P concentration [16]. The target range of dopant concentrations B and P for the feedstock qualities GEN1 and GEN2 is given in Table 3.10.

The results of vacuum purification, carried out in the 1545–1650°C range, using an industrial resistance furnace and 500 kg charges are particularly interesting. They show, in fact, significantly low silicon losses (see Table 3.11) and a very low final P and B content (see Table 3.12) that fits closely with the specifications of high quality silicon feedstocks for the P and B concentration (see Figure 3.24 [97]) and with the target values (see Table 3.10).

TABLE 3.10

Target Range of Dopant Concentrations with Purification Process GEN1 and GEN2, and Examples of the Measured Dopant Concentrations in the Feedstock

	Target Range		Feedstock Examples		
CS Silicon™	GEN1	GEN2	GEN1	GEN2 (a)	GEN2 (b)
B (ppmw)	0.15–0.4	0.1–0.2	0.19	0.16	0.11
P (ppmw)	0.8–1.1	0.1–0.6	0.88	0.45	0.1

Source: Adapted from V. Hoffmann et al., 2015. Effect of total dopant concentration on the efficiency of solar cells made in CS Silicon™, in *Proceedings of the EUVSEC 2015*, in press.

TABLE 3.11

Silicon Losses in Vacuum Purification Experiments (500 kg Samples)

Run Number	Temperature (°C)	Time (min)	Pressure (Pa)	Silicon Losses (kg)
1	1545	240	0.1–0.3	0.4
2	1545	300	0.1–0.3	0.4
3	1545	510	0.1–0.3	0.5
4	1580	180	0.1–0.3	0.5
5	1580	240	0.1–0.3	0.7
6	1580	510	0.1–0.3	2.2
7	1600	180	0.1–0.3	0.7
8	1600	240	0.1–0.3	0.7
9	1600	330	0.1–0.3	1.3
10	1600	510	0.1–0.3	2.1
11	1650	180	0.1–0.3	1.2
12	1650	240	0.1–0.3	1.6
13	1650	300	0.1–0.3	1.4
14	1650	510	0.1–0.3	2.5

TABLE 3.12

Residual Impurity Content in Vacuum Purification Experiments (500 kg Samples)

	Ca	P	B	K	Na
Starting silicon	1.4	3.3	0.2	1.1	0.5
Run number					
1	1.4	1.4	0.1	0.2	0.3
2	1.2	1.4	0.1	1.0	0.3
3	0.8	1.2	0.2	0.3	0.3
4	1.3	1.0	0.2	0.2	0.4
5	0.4	0.6	0.1	0.1	0.2
6	0.3	0.2	0.1	0.8	0.4
7	0.8	0.8	0.1	0.3	0.2
8	0.2	0.7	0.1	0.2	0.1
9	3.6	0.4	0.3	1.0	1.0
10	0.3	0.2	0.1	0.4	0.2
11	0.5	0.3	0.1	0.5	0.2
12	0.6	0.4	0.1	0.5	0.2
13	0.2	0.2	0.1	0.3	0.1
14	0.7	0.1	0.2	0.6	0.3

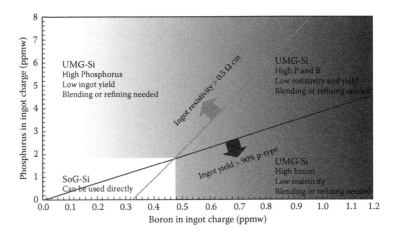

FIGURE 3.24 Useful ranges of upgraded MG-Si for PV applications: influence of the B and P concentration on the SoG quality. (After R. Tronstad. A.-K. Soiland, and E. Ennebakk, 2008. Specification of high quality silicon feedstock for solar cells, in *Proceedings of the Workshop Crystal Clear: Arriving at Well-Founded SoG Silicon Feedstock Specifications*, Amsterdam, the Netherlands. Permission obtained from Elkem, January 22, 2016.)

The low doping and compensation levels of the GEN2 material can be correlated with high minority carrier mobility and high solar cell efficiencies [96].

The results of many different industrial tests with current CS silicon at different stages of the PV chain showed, in fact, efficiencies of 17.3%, 0.1% below the reference cells [16].

In a comparative study between typical mc-Si and CS-Si-based modules, evaluated in real outdoor conditions during a full-year period, no difference was found between both material types [98].

The FerroSolar process is easily applicable to industrial scale, its results are reproducible, and it is cheaper and more environmentally friendly than the Siemens process.

3.3.1.2   The Photosil Project

The Photosil process was developed by FerroPem (part of the Ferroatlantica group) in 2004, in its Photosil site located in Chambery [99]. It aimed at producing SoG-Si, able to replace directly (without blending) the chunk material from chlorosilane processes. The Photosil silicon can be used without additional dopants, in standard solar cell production factories.

The process (see flowsheet in Figure 3.25) begins by a careful selection of the raw materials (silica, carbon-rich reducers) used to produce MG-Si in classic reduction furnaces. Then, two segregation steps are used to remove metallic impurities and a part of phosphorus, giving upgraded metallurgical silicon (UMG-Si) which is treated with plasma to remove boron and undergoes a final segregation to remove carbon and the remaining metals and phosphorus.

The innovative part of the Photosil process is the plasma purification step, developed in the frame of a research-industry consortium with Apollon Solar, CEA, and

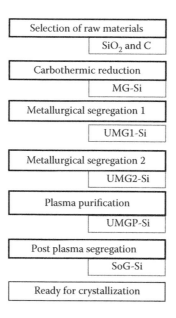

FIGURE 3.25   Flowsheet of the Photosil refining process.

at the SIMaP laboratory of CNRS in Grenoble [37], where 10 kg of silicon were already successfully processed.

The Photosil process started by reducing the boron concentration of 75 kg batches down to 3.5 ppmw in 2005 [100] and was up-scaled to 250 kg batches purified at 0.3 ppmw of boron in 2012 [99]. This evolution was mainly cost driven, with a second-generation plasma facility announced in 2011 [101], and running since 2013, with an objective cost below \$15/kg for the whole production chain of SoG-Si, vertically integrated from the carboreduction of silica up to Si blocks ready for crystallization.

Other parts of the process include segregation steps (before and after the plasma), and upstream optimization of the metallurgical silicon. These steps were optimized in parallel with the plasma process, to reach the SEMI specifications for SoG-Si (including a phosphorus concentration below 0.6 ppmw), together with quality criteria defined from the consortium's experience, such as a resistivity between 0.6 and 3 Ω cm [99]. A dedicated study of the post plasma segregation is reported in Reference 102.

Photosil material has been used since 2006 to produce mc-Si ingots and since 2012 to produce single crystal boules by the CZ technique.

Table 3.13 [99–101,103–107] reports details about the evolution of the Photosil quality and its use along the development period. The feedstock used was simply a plasma purified material in early studies, then the standard silicon (Std) produced by the Photosil process, eventually trimmed to get n-type silicon (n-Std). Eventually, an optimized material (Opt or n-Opt), better purified, was used for studies on high performance cells. The impurity contents of Std and Opt feedstocks evolved during the development, showing that the process can be adjusted to meet various purity requirements.

TABLE 3.13
Photosil Material Composition for Crystallized Samples

Year and Material		Impurity Contents (ppmw)					Crystallization	References
		B	P	Al	Fe	Others		
2006	UMGP	3.5	2.5	5		Ca = 9, C = 15–20, O = 20–25	mc 10 kg	[100]
2008	UMGP	1.5	2	0.2		Cu = 0.17, Zn = 0.26	mc 8 kg	[103]
2010	Std	1.5	4	<2	<5	Cu < 2, Ti < 1	mc 10 × 40 kg	[104]
	Opt	0.3	1	<2	<2	Cu < 2, Ti < 1	mc 2 × 40 kg	
2011	Std	0.5	1.1	0.1	0.05	Cu = 0.15	mc 3 × 400 kg	[101]
2012	Std	0.32	0.8				mc 35 kg	[105]
	Std-n	0.3	2				CZ 15 kg	[106]
2013	Std	0.3	0.6				CZ 10–15 kg	[107]
	Opt	0.12	0.4					
	Opt-n	0.12	0.8					
	Std	<0.3	<0.6	<0.1	<0.1	Cu < 0.1	mc 2 × 440 kg	[99]

TABLE 3.14
Reported Cell Performance with 100% Photosil Material

Year and Material		Cell Process	No. of Cells	Mean	Max	LID (rel%)	References
				Cell Efficiency			
2010	Std-mc	Std indus Photowatt	3000	15%	16%		[102]
	Opt-mc	Screen-printed National Solar Energy Institute (INES)	~10	15.5%	16.2%		[108]
	Opt-CZ	Screen-printed INES	~20	17.4%	17.6%	3	[109]
2011	Std-mc	Std indus Pvalliance mean = 16% w.EG-Si	16 × 54	15.5%	16.4%		[101]
	Opt-mc	Screen-printed INES		16.3%	16.7%	<2	
2012	Std-CZ	Screen-printed INES	~20	17.5%	18%	12	[105]
	Std-nCZ	Hetero Roth & Rau	14	18.6%	19%		[106]
2013	Std-CZ	Screen-printed INES	10	17.9%	18.6%	2.2	[107]
	Opt-CZ		10	18.1%	18.2%	1.8	
	Std-nCZ	Bifacial screen INES mean = 18.7% w.EG-n	10	18%	18.3%	1	
	Opt-nCZ		10	18.5%	18.8%	0.6	
	Std-mc	Std indus Irysolar	5500	16.1%	16.9%		[99]

The crystallization experiments on Photosil material (see again Table 3.13) were carried out at the laboratory scale, except in 2011 and 2013 when half-ton-sized ingots were grown to be delivered to solar cell manufacturers. The material grown in 2011 was used for the production of 225 W-modules for a 100 kW solar farm, where 80 Photosil modules were monitored together with standard modules made out of EG-Si [99]. The power output of Photosil and standard modules was shown to be very close.

The material grown later was delivered to different solar cell manufacturers to test the efficiency and the light-induced degradation (LID) effects. The results of these tests are reported in Table 3.14.

The cells made with Photosil material reached comparable efficiency (not reported in Table 3.14) of cells manufactured from EG-Si for each kind of cell manufacturing process, including high performance monocrystalline cells, for which an efficiency up to 19% was measured.

The LID has been reduced below 2% of the initial efficiency, and other criteria needed for industrial cell production (high ingot yield, low percentage of broken wafers, or off-spec cells) have been reached as well.

The material is now available for further industrial evaluation.

3.3.2 Silicor Materials Process for Silicon Purification

Silicor Materials was founded as an R&D company in 2006 under the name Calisolar, with the goal of manufacturing low-cost PV solar cells from upgraded MG-Si. After merging with the Canadian company 6N Silicon, it had its own internal silicon purification process which was optimized for the requirements of a solar cell production. Today, the company is fully focused on the metallurgical route of silicon purification and intends to first build a large plant of up to 19,000 t annual capacity in Grundartang in Iceland.

The Silicor Materials process can be described as a liquid to solid refining process comprising a number of refining steps, whereby, MG-Si with a purity level of at least 99.5% Si is refined up to the levels needed for solar application. The four principal refining steps, as also shown in Figure 3.26 are:

- *Solvent growth refining.* A molten aluminum–silicon alloy is allowed to cool, leading to silicon refining in the form of aluminum-coated silicon crystals.
- *Wet-chemical aluminum removal.* The aluminum coating on the silicon crystals is dissolved in hydrochloric acid forming polyaluminum chloride (PAC).
- *DS.* Clean silicon crystals are re-melted and directionally solidified resulting in fully refined silicon ingots.
- *Final preparation.* Purified silicon ingots are cleaned, cut, chunked, and blended to eliminate chemical variation.

In addition to the primary output of SoG-Si, the process route has two by-products of high commercial value, that is, an Al–Si master alloy for use in the aluminum casting industry and PAC, used for waste-water treatment applications.

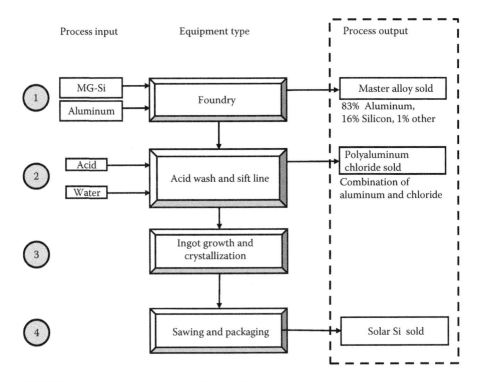

FIGURE 3.26 Map of process flow for the silicor process.

3.3.2.1 Solvent Growth Refining

A molten aluminum–silicon mix, containing approximately 40% silicon by weight (Al–40 Si wt%) is the starting point for the Silicor Materials process. This is accomplished by combining, during the solvent growth refining step, MG-Si (99.57% Si) with aluminum or a frozen eutectic coming from the process and containing a high amount of Al.

The full phase diagram for the Al–Si system is given in Figure 3.27, which also shows a visual representation of the typical solidification pathway for a slowly cooled melt. With any hypereutectic Al–Si melt where the weight percent of Si is greater than 12.6%, slow cooling from the liquid state will form silicon crystals of high purity to solidify first and will cause an enhanced segregation effect due to the low temperatures of this crystallization. As more silicon crystals solidify, the remaining melt will experience an increase in the concentration of aluminum and other elements, compared to the composition of the original melt.

In such a melt that exhibits a nondirectional temperature gradient, the silicon will grow from the mold walls and in the form of platelets or agglomerates of platelets called "flakes," resulting in a three-dimensional network of silicon flakes with the remaining melt in its cavities.

Figure 3.28 shows a scheme of the solvent refining step. Here, a counter-current system is used to achieve purification over a number of passes, which involve the melting and solidification of the Al–Si alloy at ever increasing purity. In effect, the incoming pure aluminum acts as a diluent of the impurity levels and the solidification at each pass acts to segregate these impurities so that they end up in the master aluminum alloy.

In such a process the clean silicon, which already has seen several passes of purification, can only get in contact with the purest aluminum making sure there is no cross contamination between different levels of the purification steps. The number of passes can be varied according to the quality of the incoming material and the target specification of the process [110].

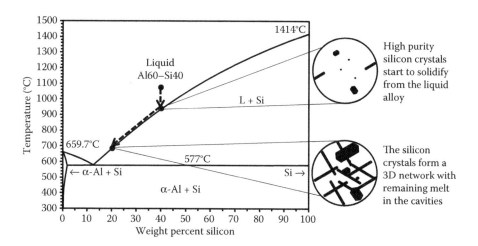

FIGURE 3.27 Phase diagram of the Al–Si system.

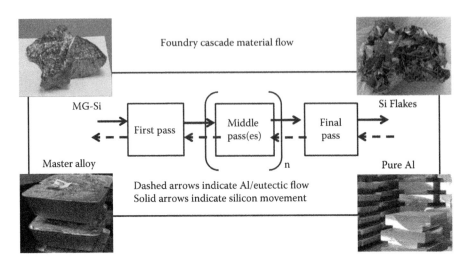

FIGURE 3.28 Material flow during the solvent refining.

After a sequence of passes, when the flake purity meets the process specification, the grown flakes are separated from the remaining eutectic. In every pass, this has to be done at a temperature above the eutectic point (577°C) to be able to drain the remaining melt from the Si network. Usually at this stage, the network is stable enough to stay intact during draining but also has enough porosity to let the melt go. Afterward silicon flakes with only a bit of aluminum coating can be harvested. The master alloy coming out of this process is poured into molds to be sold to the aluminum casting industry.

3.3.2.2 Wet-Chemical Aluminum Removal

The next principal process step is an acid wash step, which removes the aluminum present on the surface of the silicon crystals. Screening (sifting) takes place after the acid wash to remove nonstandard Si-flakes. In addition to creating clean flakes for the DS process, the acid washing process also creates PAC from the dissolution of aluminum in the hydrochloric acid. This by-product can be used as a flocculant during waste-water treatment applications.

The hydrogen evolution during the etching reaction is kept below the lowest explosion limit by air circulation. With constant process control in the etching section, using a proper HCl addition, a peak hydrogen formation is prevented.

It is targeted that the aluminum content of the etched flakes will be limited to 2500 ppmw. This value is higher than the maximum solubility of Al in Si (412 ppmw) [111] since some of the aluminum present in the flakes is locked in the silicon flake structures and cannot be digested by the hydrochloric acid.

The etched and rinsed flakes are then dried to be ready to be melted in a DS furnace.

3.3.2.3 DS and Final Preparation

The DS process consists of melting the flakes by using an induction furnace followed by casting and DS in molds. This step is primarily responsible for the final removal of Al from the silicon flakes that arrive from the acid washing process. There is reduction of other impurities, but these have already been significantly reduced in the solvent growth refining step. During DS, metals segregate readily to the top of the ingot where they can physically be removed once the ingot is solid. This top material is then recycled back to the solvent refining step. Since the Al–Si purification process has multiple levels of purity over each step, it also is well suited to recycle silicon of any grade. Depending on the purity of the recycled material, it can go the shortest possible loop through the purification without wasting energy. This makes the process also potentially interesting for future tasks in the solar industry when more material, for example, from old solar cells and other processes, needs to be recycled.

Figure 3.29 shows the principal elements of the DS mold, where a top heater and air flow at the bottom of the mold ensure that the temperature gradient runs from top to bottom of the mold geometry.

The removal of Al during this process stage is further assisted by the use of a pre-fused flux both in the furnace and applied in a thin layer on the surface of the melt during DS. This flux is mentioned as the "cover flux" in Reference 68 and is a mixture of three components, SiO_2, Na_2CO_3, and CaO. The oxidation potential of this mixture is such to favor the preferential oxidation of Al and results in aluminum oxide (see Section 3.2.2), which will be dissolved by the liquid glass covering the melt.

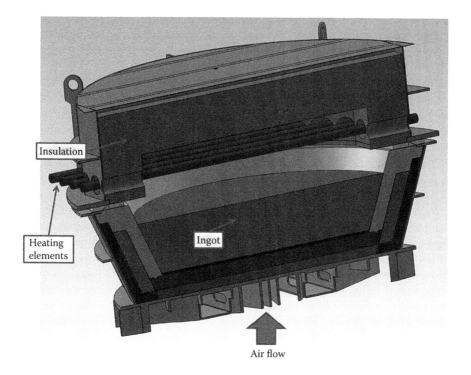

Air flow

FIGURE 3.29   DS mold.

Using a pure sodium silicate glass, which is highly efficient for the Al oxidation, the feedstock obtained after a solidification refining presents an Al content around 0.1 ppmw, and could be used with success as a solar silicon feedstock [69,112].

The final processing involves the cleaning and cutting of the ingots, a thermal cracking process (chunking), and finally mixing of batches to reduce chemical variation if necessary, followed by packing and labeling.

3.4 CONCLUSIONS

The main result of R&D activities concerning MG-Si purification is that a physical limit to its full up-grading to SoG quality exists, since even the most advanced processes are unable to bring the concentration of dopant impurities (B and P) down to that of EG-Si. The removal of other impurities is, instead, physically and experimentally feasible. The product of MG-Si purification is, therefore, a compensated material, which nevertheless is demonstrated to be of possible use for solar cell production, with limited losses of efficiency in comparison with cells manufactured with EG-Si.

As the high B and P contamination of MG-Si depends only on the B content of quartz used as the silicon source and of the P content of the reductants, which are used to feed the carboreduction furnace (see Chapter 2), the only solution possible to get a pure, undoped silicon would be the use of high-grade quartz and P-free reductants.

Cost problems concerning B-loan quartz and difficult furnace operation with nonconventional reductants already suggested to Elkem (see Chapter 4) to neglect the high purity option, which remains, however, open for further progress in terms of operational capability of the carbothermic furnace, possibly with pure, synthetic SiO_2.

It should be noted that the impressive progress in non-dopant impurity refining processes is eminently due to the intelligent application of thermodynamic principles and of material science schemes, which also open up hope for further progress in this respect, as seem to be shown by the Solsilc and the Silicor projects, which translate these schemes to industrial applications.

The problem, however, remains that a long sequence of different chemical and physical steps is necessary to obtain the required purity.

Electrochemistry could offer viable options for single-step purification processes, although it is a matter still insufficiently studied. Here, the challenge is either to optimize the electrowinning processes or those based on the use of liquid silicon alloys at the cathode. The efficiency of the electrowinning process depends on the diffusivity of the impurities and could envisage a preliminary electroless step for the removal of most electronegative impurities.

The use of liquid alloys as the cathode helps to suppress the large overvoltage effects arising at solid silicon cathodes and to work at temperatures lower than the melting point of silicon, with the advantage to reduce or even suppress the compatibility problems with fluoride melts. Eventually, silicon alloys with a volatile or an acid-soluble second component should be considered with attention, taking into account the already existing experimental knowledge of sublimation and leaching processes of silicon alloys.

REFERENCES

1. B. Ceccaroli and S. Pizzini, 2012. Processes, in *Advanced Silicon Materials and Processes*, S. Pizzini (Ed.), John Wiley & Sons, Chichester, UK, pp. 21–78.
2. N. Yuge, M. Abe, K. Hanazawa, H. Baba, N. Nakamura, Y. Kato, Y. Sakaguchi, S. Hiwasa, and F. Aratani, 2001. Purification of metallurgical-grade silicon up to solar grade. *Prog. Photovolt. Res. Appl.*, 9, 203–209.
3. P. Woditsch and W. Koch, 2002. Solar grade silicon feedstock supply for PV industry, *Sol. Energy Mater. Sol. Cells*, 72, 11–26.
4. E. J. Øvrelid, K. Tang, T. Engh, and M. Tangstad, 2009. Feedstock, in *Crystal Growth of Silicon for Solar Cells*, K. Nakajima, N. Usami (Eds.), Springer, Berlin Heidelberg, pp. 1–23, ISBN 978-3-642-02044-5.
5. A. Ciftja, Th. Abelengh, and M. Tangstad, 2008. *Refining and Recycling of Silicon: A Review*, Norwegian University of Science and Technology, Trondheim, February.
6. M. D. Johnston, L. T. Khajavi, M. Li, S. Sokhanvaran, and M. Barati, 2012. High-temperature refining of metallurgical-grade silicon: A review, *JOM J. Miner. Met. Mater. Soc.*, 64, 935–945.
7. J. Safarian, G. Tranell, and M. Tangstad, 2012. Processes for upgrading metallurgical grade silicon to solar grade silicon, *Energy Procedia*, 20, 88–97.
8. Y. Delannoy, 2012. Purification of silicon for photovoltaic applications, *J. Cryst. Growth*, 360, 61–67.
9. H. Sasaki, Y. Kobashi, T. Nagai, and M. Maeda, 2013. Application of electron beam melting to the removal of phosphorous from silicon: Toward production of solar grade silicon by metallurgical processes, *Adv. Mater. Sci. Eng.*, 2013, 8 pp., Article ID 857196, http://dx.doi.org/10.1155/2013/857196.
10. M. Heuer, 2013. Metallurgical grade and metallurgically refined silicon for photovoltaics, in *Semiconductors and Semimetals*, G. P. Willeke and E. Weber (Eds.), Vol. 89, Elsevier B.V., pp. 77–134.
11. A. K. Søiland, M. G. Dolmen, J. Heide, U. Thisted, G. Halvorsen, G. Ausland, K. Friestad et al., 2014. First results from a simplified Elkem solar route—Input to tolerance limits, *Sol. Energy Mater. Sol. Cells*, 130, 661–667.
12. S. Pizzini, 2015. *Physical Chemistry of Semiconductor Materials and Processes*, John Wiley & Sons, Chichester, UK, pp. 274–279.
13. J. Safarian and M. Tangstad, 2012. Vacuum refining of molten silicon, *Metall. Mater. Trans. B*, 43, 1427–1445.
14. R. Suzuki, K. Sakaguki, T. Nakagiri, and N. Sano, 1990. Gaseous removal of phosphorous and boron from molten silicon (in Japanese), *J. Jpn. Inst. Met.*, 54, 161–167.
15. S. Zheng, T. A. Engh, M. Tangstad, and X. I. Luo, 2011. Numerical simulations of phosphorous removal from silicon by induction vacuum refining, *Metall Mater. Trans. A*, 42a, 2214–2225.
16. A. Souto, J. Bullon, R. Ordas, and J. M. Migues, 2014. Industrial scale vacuum applications in the FerroSolar project, in *Silicon for the Chemical and Solar Industry XII*, Trondheim, Norway, June 23–26, NTU Publ., pp. 67–76.
17. C. P. Khattak, D. B. Joyce, and F. Schmid, 2008. Development of solar grade (SoG) silicon. *Final Report for DOE SBIR Phase II, Contract Number DE-FG02-04ER83928*, January 16.
18. J. Safarian and M. Tangstad, 2010. Vacuum behaviour of the dissolved elements in molten silicon, in *Silicon for the Chemical and Solar Industry X*, Alesund-Geiranger, Norway, June 28–July 2, NTU Publ., pp. 41–49.
19. J. J. Wu, W.-H. Ma, B. Yang, Y.-N. Dai, and K. Morita, 2009. Boron removal from metallurgical grade silicon by oxidizing refining, *Trans. Nonferr. Met. Soc. China*, 19, 463–467.

20. W. Ji, W.-H. J. Wu, Y. Li, W. Ma, K. Liu, K. Wei, K. Xie, B. Yang, and Y. Dai, 2014. Impurity removal from metallurgical grade silicon using gas blowing refining techniques, *Silicon*, 6, 79–85.
21. I. Kero, M. K. Ness, V. Andersen, and G. M. Tranell, 2015. Refining kinetics of selected elements in the industrial silicon process, *Metall. Mater. Trans.*, 46B, 1186–1194.
22. M. Bolandi, 2012. Refining metallurgical grade silicon by chlorination treatment with emphasis on aluminum removal, Master of Applied Science thesis, School of Graduate Studies, McMaster University, Hamilton, Canada.
23. Y. Shen, 2015. Carbothermal synthesis of metal-functionalized nanostructures for energy and environmental applications, *J. Mater. Chem. A*, 3, 13114–13118.
24. J. K. Tuset, 1992. The refining of silicon and ferrosilicon, in *Proceedings of the International Ferroalloys Congress*, SAIMM, Johannesburg, pp. 193–199.
25. J. Smith, 1997. Silicon refining process, Patent WO 1998011018 A1, August 27.
26. S. M. Schnurre, J. Grobner, R. Schmid-Fetzer, 2004. Thermodynamics and phase stability in the Si–O system, *J. Non-Cryst. Sol.*, 336, 1–25.
27. A. Muhlbauer, V. Diers, A. Walter, and J. G. Grabmaier, 1991. Removal of C/SiC from liquid silicon by directional solidification, *J. Cryst. Growth*, 108, 41–52.
28. A. K. Sakaguchi and M. Maeda, 1992. Decarburization of silicon melt for solar cells by filtration and oxidation, *Metall. Trans. B*, 23B, 423–427.
29. A. K. Søiland, 2004. Silicon for solar cells, Thesis, NTNU, Trondheim, Norway.
30. J. Hintermayer, 2011. Process for decarburization of a silicon melt WO 2011088953 A1 28/07/2011 assigned to Evonik, Degussa.
31. C. P. Khattak, D. B. Joyce, and F. Schmid, 2000. Upgrading metallurgical grade (MG) silicon for use as a solar grade feedstock, in *Proceedings of the 28th IEEE PV Specialist Conference,* Anchorage, Alaska, pp. 49–52.
32. C. P. Khattak, D. B. Joyce, and F. Schmid, 2001. Production of solar grade (SoG) silicon by refining liquid metallurgical grade (MG) silicon, *Final Report NREL/SR-520-30716*, Crystal Systems, Inc., April 19.
33. C. P. Khattak, D. B. Joyce, and F. Schmid, 2002. A simple process to remove boron from MG-grade silicon, *Sol. Energy Mater. Sol. Cells*, 74, 77–89.
34. E. F. Nostrand and M. Tangstad, 2012. Removal of boron from silicon by moist hydrogen gas, *Metall. Mater. Trans. B*, 43B, 814–822.
35. J. Wu, Y. Li, W. Ma, K. Lin, K. Wei, K. Xie, B. Yang, and Y. Dai, 2014. Impurity removal from metallurgical grade silicon using gas blowing refining techniques, *Silicon*, 6, 79–85.
36. N. Nakamura, H. Baba, Y. Sakaguchi, and Y. Kato, 2004. Boron removal in molten silicon by a steam-added plasma melting method, *Mater. Trans.*, 45, 858–864.
37. C. Alemany, C. Trassy, B. Pateyron, K. I. Li, and Y. Delannoy, 2002. Refining of metallurgical-grade silicon by inductive plasma, *Sol. Energy. Mater. Sol. Cells*, 72, 41–48.
38. D. Morvan, J. Amouroux, M. C. Charpin, and H. Lauvray, 1983. Elimination du bore dans le silicium par fusion de zone sous plasma inductif haute fréquence:rôle des plasmas réactifs et du laitier. Caractérisation du silicium photovoltaïque, *Rev. Phys. Appl. (Paris)*, 18, 239–251.
39. K. Suzuki, T. Kumagai, and N. Sano, 1992. Removal of boron from metallurgical grade silicon by applying the plasma treatment, *ISIJ Int.*, 32, 630–634.
40. T. Ikeda and M. Maeda, 1996. Purification of metallurgical silicon for solar-grade silicon by reactive rotating plasma arc melting, *Mater. Trans. JIM*, 37, 983–987.
41. J. Altenberend, 2012. Kinetics of the plasma refining process of silicon for solar cells: Experimental study with spectroscopy, PhD thesis, Grenoble University, Saint-Martin-d'Hères, France.
42. J. Altenberend, G. Chichignoud, and Y. Delannoy, 2013. Atomic emission spectroscopy method for mixing studies in high power thermal plasmas, *Spectrochim. Acta B*, 89, 93–102.

43. K. Tang, S. Andersson, E. Nordstrand, and M. Tangstad, 2012. Removal of boron in silicon by H_2/H_2O gas mixtures, *JOM*, 64, 952–956.
44. O. S. Sortland and M. Tangstad, 2014. Boron removal from silicon melts by H_2O/H_2 gas blowing: Mass transfer in gas and melt, *Metall. Mater. Trans.*, E 1, 211–225.
45. B. P. Lee, H. M. Lee, D. H. Park, J. S. Shin, T. U. Yu, and B. M. Moon, 2011. Refining of MG-Si by hybrid melting using steam plasma and EMC, *Sol. Energy Mater. Sol. Cells*, 95, 56–58.
46. W. R. Imler, R. E. Haun, R. A. Lampson, M. Charles, and P. Meese, 2013. Efficacy of plasma arc treatment for the reduction of boron in the refining of solar-grade silicon, in *Proceedings of the 39th IEEE PVSC*, Tampa, Florida, June 16–21, pp. 497–501.
47. M. Ratto, E. Ricci, E. Arato, and P. Costa, 2001. Oxidation of metals with highly reactive vapors: Extension of Wagner theory, *Metall. Mater. Trans.*, B 32, 903–911.
48. Y. Delannoy, G. Chichignoud, and M. Vadon, 2014. Kinetic model for gas–liquid extraction of boron from solar silicon: The role of H_2, *Presented at 143rd Annual TMS Meeting*, San Diego, California.
49. NIST Reference Data Base IIA1(XVA).
50. R. Noguchi, K. Susuki, T. Tsukihashi, and N. Sano, 1994. Activity coefficients of B in Si, *Metall. Mater. Trans.*, 25B, 903–907.
51. Y. Delannoy, C. Alemany, K.-I. Li, P. Proulx, and C. Trassy, 2002. Plasma refining process to provide solar grade silicon, *Sol. Energy Mater. Sol. Cells*, 72, 69–75.
52. O. S. Sortland and M. Tangstad, 2014. Boron removal in silicon by H_2/H_2O gas mixtures, *JOM*, 64, 952–956.
53. G. Halvorsen, 1984. Method for production of pure silicon, US Patent 4539194 A.
54. A. Schei, 1985. High purity silicon production, in *Refining and Alloying of Liquid Aluminium and Ferro-Alloys*, T. A. Engh, S. Lyng, and H. A. Øye (Eds.), Aluminium-Verlag, Trondheim, Norway; A. Schei, 1986. Metallurgical production of high purity silicon, in *INFACON 86 Proceedings*, Rio de Janeiro, Brazil, August 31–September 3, pp. 389–398.
55. L. K. Jacobsen, 2013. Distribution of boron between silicon and $CaO–SiO_2$, $MgO–SiO_2$, $CaO–MgO–SiO_2$ and $CaO–Al_2O_3–SiO_2$ slags at 1600°C, PhD thesis, NTNU, ISBN 978-82-471-4791-7 (electronic version).
56. K. Suzuki, T. Sugiyama, K. Takano, and N. Sano, 1990. Thermodynamics for removal of boron from metallurgical silicon by flux treatment, *J. Jpn. Inst. Met.*, 54, 168–172.
57. M. Tanahashi, Y. Shimpo, T. Fujisawa, and C. Yamauchi, 2002. Distribution behavior of boron between SiO_2 saturated $Na_2O–CaO–SiO_2$ flux and molten silicon, *J. Min. Mater. Process. Inst. Jpn.*, 118, 497–505.
58. L. Pelosini, A. Parisi, and S. Pizzini, 1980. Process for purifying silicon, Patent 4241037, December 23.
59. T. A. Engh, 2002. *Principles of Metal Refining*, Oxford Science Publications, Oxford, UK.
60. J. L. Murray and A. J. McAlister, 1984. The Al–Si (aluminum–silicon) system, *Bull. Alloy Phase Diagr.*, 5(1), 74–84.
61. T. Yoshikawa and K. Morita, 2005. Refining of silicon by the solidification of Si–Al melts with electromagnetic force, *ISIJ Int.*, 45, 967–971.
62. K. Morita and T. Yoshikawa, 2008. Low temperature refining of solar grade silicon using Si–Al solvent, in *Silicon for the Chemical and Solar Industry IX*, Oslo, Norway, June 23–26, NTU Pub., pp. 41–59.
63. T. Yoshikawa and K. Morita, 2009. Refining of silicon during its solidification from a Si–Al melt, *J. Crys. Growth*, 311, 776–779.
64. K. Morita and T. Yoshikawa, 2011. Thermodynamic evaluation of new metallurgical refining processes for SOG-silicon production, *Trans. Nonferr. Met. Soc. China*, 21, 685–690.
65. M. Rodot, J. E. Bourree, A. Mesli, G. Revel, R. Kishore, and S. Pizzini, 1987. Al-related recombination center in polycrystalline silicon, *J. Appl. Phys.*, 62, 2556–2558.

66. S. Rein, 1999. ISE Freiburg (unpublished results).
67. G. Coletti, D. MacDonald, and D. Yang, 2012. Role of impurities in solar silicon, in *Advanced Silicon Materials for Photovoltaic Applications*, S. Pizzini (Ed.), John Wiley & Sons, Chichester, UK, pp. 79–125.
68. A. Turenne, A. Nouri, and C. A. Christian, 2013. Cover flux and method for silicon purification, Patent Application WO 2014118630 A1.
69. M. Heuer, A. Turenne, M. Kas, F. Kirscht, and T. Jester, 2015. Status and perspectives of UMG silicon, *Presented at CSSC-8*, Bamberg, Germany, May 5–8.
70. J. Cai, X.-T. Luo, G. M. Haarberg, O. E. Kongstein, and S.-L.Wang, 2012. Electrorefining of metallurgical silicon in molten $CaCl_2$ based melts, *J. Electrochem. Soc.*, 159, D155–D158.
71. S. Pizzini and R. Morlotti, 1965. Oxygen and hydrogen electrodes in molten fluorides, *Electrochim. Acta*, 10, 1033–1039.
72. S. Fabre, C. Cabet, L. Cassayre, P. Chamelot, S. Delepech, J. Finne, L. Massot, and D. Noel, 2013. Use of electrochemical techniques to study the corrosion of metals in model fluoride melts, *J. Nucl. Mater.*, 441, 583–591.
73. E. Olsen and S. Rolseth, 2010. Three-layer electrorefining of silicon, *Metall. Mater. Trans.*, 41B, 295–302.
74. F. Grjotheim, K. Matisovsky, and P. Fellner, 1973. Some aspects of electrochemical preparation of silicon alloys, *Chem. Zvesti*, 27, 165–171.
75. R. Monnier and J. C. Giacometti, 1964. Researches sur le raffinage electrolytique de silicium, *Helv. Chim. Acta*, 47, 345–353.
76. D. Elwell and G. M. Rao, 1988. Electrolytic production of silicon, *J. Appl. Electrochem.*, 18, 15–22.
77. S. V. Kuznetsova, V. S. Dolmatov, and S. A. Kuznetsov, 2009. Voltammetric study of electroreduction of silicon complexes in a chloride–fluoride melt, *Russ. J. Electrochem.*, 54, 742–748.
78. U. Cohen and R. A. Huggins, 1976. Silicon epitaxial growth by electrodeposition from molten fluorides, *J. Electrochem. Soc.*, 123, 381–383.
79. R. C. DeMattei and R. S. Feigelson, 1978. Growth rate limitations in electrochemical crystallization, *J. Cryst. Growth*, 44, 115–120.
80. G. M. Rao, D. Elwell, and R. S. Feigelson, 1980. Electrowinning of silicon from K_2SiF_6–molten fluoride systems, *J. Electrochem. Soc.*, 127, 1940–1944.
81. G. M. Rao, D. Elwell, and R. S. Feigelson, 1981. Electrodeposition of silicon onto graphite, *J. Electrochem. Soc.*, 128, 1708–1712.
82. T. Nohira, K. Yasuda, and Y. Ito, 2003. Pinpoint and bulk electrochemical reduction of insulating silicon dioxide to silicon, *Nature Mater.*, 2, 397–401.
83. K. Yasuda, T. Nohira, and Y. Ito, 2005. Effect of electrolysis potential on reduction of solid silicon dioxide in molten $CaCl_2$, *J. Phys. Chem. Solids*, 66, 443–447.
84. K. Yasuda, T. Nohira, R. Hagiwara, and Y. H. Ogata, 2007. Direct electrolytic reduction of solid SiO_2 in molten $CaCl_2$ for the production of solar grade silicon, *Electrochim. Acta*, 53, 106–110.
85. E. Olsen, 2004. Electrolyte and method for manufacturing and/or refining of silicon, US Patent Application 2004/0238372A1, December 2.
86. S. V. Devyatkin, 2003. Electrochemistry of silicon in chloro-fluoride and carbonate melts, *J. Min. Met.*, 39, 303–307.
87. V. I. Shapoval, V. V. Malyashev, I. A. Novoselova, and Kh. Kushkhov, 1995. Modern problems in the high temperature electrochemical synthesis of the compounds of Group IV–VI transition metals, *Russ. Chem. Rev.*, 64, 125–132.
88. E. Stefanidaki, G. M. Photiadis, C. G. Kontoyannis, A. F. Vik, and T. Ostvold, 2002. Oxide solubility and Raman spectra of NdF_3, $LiF–KF–MgF_2–Nd_2O_3$ melts, *J. Chem. Soc. Dalton Trans.*, 11, 2302–2307.

89. M. D. Johnston, L. T. Khajavi, M. Li, S. Sokhanvaran, and M. Barati, 2012. High temperature refining of metallurgical-grade silicon: A review, *JOM*, 64, 935–945.

90. J. M. Olson and K. L. Carleton, 1981. A semipermeable anode for silicon electrorefining, *J. Electrochem. Soc.*, 128, 2698–2699.

91. E. Olsen, S. Rolseth, and J. Thonstad, 2010. Electrocatalytic formation and inactivation of intermetallic compounds in electrorefining of silicon, *Metall. Mater. Trans.*, 41B, 752–757.

92. E. Øvrelid, B. Geerlichs, A. Waernes, O. Raaness, I. Solheim, R. Jensen, K. Tang, S. Santeen, and B. Wiersma, 2006. Solar grade silicon by a direct metallurgical process, in *Proceedings of Silicon for the Chemical Industry VIII*, Trondheim, Norway, NTU Publ., pp. 223–234, June 12–16.

93. M. Jia, Y. Cheng, Z. Tian, Y. Lai, and Y. Liu, 2014. Electrowinning of silicon with liquid electrodes, in *Proceedings of the EDP Congress*, San Diego, California, February 16–20, pp. 353–360.

94. J. Bullon, R. Ordas, Th. Margaria, A. Miranda. J. M. Miguez, A. Perez, and A. Souto, 2010. Ferrosolar project, situation and perpectives, in *Proceedings of Silicon for Chemical & Solar Industry X*, Alesund-Geiranger, Norway, NTU Publ., pp. 179–189.

95. J. Bullon, 2014. The application of vacuum in Ferrosolar project, *Presented at Silicon for Chemical & Solar Industry XII*, Trondheim, Norway, June 23–26.

96. V. Hoffmann, J. M. Miguez, J. Bullon, I. Buchowska, T. Vlasenko, S. Beringov, J. Denafas, K. Meskeriviciene, L. Petrieniene, and R. Ordas, 2015. Effect of total dopant concentration on the efficiency of solar cells made in CS Silicon™, in *Proceedings of the 31th EUPVSEC*, Hamburg, Germany, September 14–18, pp. 559–563.

97. R. Tronstad, A.-K. Soiland, and E. Ennebakk, 2008. Specification of high quality silicon feedstock for solar cells, Presented at *the Workshop Crystal Clear: Arriving at Well-Founded SoG Silicon Feedstock Specifications*, Amsterdam, The Netherlands, November 13–14.

98. E. Sanchez, J. Torreblanca, I. Guerrero, T. Carballo, V. Parra, R. Ordas, J. Bullon et al., 2011. Evaluation of performance of standard and UMG multicrystalline silicon modules in outdoor conditions, in *Proceedings of the 26th EUPVSEC*, Hamburg, Germany, pp. 3657–3660, ISBN 3-936338-27-2.

99. F. Cocco, D. Grosset-Bourbange, P. Rivat, G. Quost, J. Degoulange, R. Einhaus, M. Forster, and H. Colin, 2013. Photosil UMG silicon: Industrial evaluation by multi-c p-type ingots and solar cells, in *Proceedings of the 28th EUPVSEC*, Paris, France, pp. 1435–1438.

100. R. Einhaus, J. Kraiem, F. Cocco, Y. Caratini, D. Bernou, D. Sarti, G. Rey et al., 2006. PHOTOSIL—Simplified production of solar silicon from metallurgical silicon, in *Proceedings of the 21st EUPVSEC*, Dresden, Germany, pp. 580–583.

101. T. Margaria, F. Cocco, L. Neulat, J. Kraiem, R. Einhaus, J. Degoulange, D. Pelletier et al., 2011. UMG silicon from the PHOTOSIL project—A status overview in 2011 on the way towards industrial production, in *Proceedings of the 26th EUPVSEC*, Hamburg, Germany, pp. 1806–1809.

102. B. Drevet, D. Camel, N. Enjalbert, J. Veirman, J. Kraiem, F. Cocco, E. Flahaut, and I. Périchaud, 2010. Segregation and crystallization of purified metallurgical grade silicon: Influence of process parameters on yield and solar cell efficiency, in *Proceedings of the 25th EUPVSEC/5th WCPEC*, Valencia, Spain, pp. 1238–1243.

103. J. Degoulange, I. Perichaud, C. Trassy, and S. Martinuzzi, 2008. Multicrystalline silicon wafers prepared from upgraded metallurgical feedstock, *Sol. Energy Mater. Sol. Cells*, 92, 1269–1273.

104. J. Kraiem, F. Cocco, B. Drevet, N. Enjalbert, S. Dubois, D. Camel, D. Grosset-Bourbange, R. Einhaus, D. Pelletier, and T. Margaria, 2010. High performance solar cells made from 100% UMG silicon obtained via the Photosil process, in *Proceedings of the 35th IEEEPVSC*, Honolulu, Hawaii, June 20–25, pp. 001427–001431, DOI: 10.1109/PVSC.2010.5614418.
105. J. Degoulange, N. Enjalbert, F. Cocco, J. Kraiem, D. Grosset-Bourbange, S. Dubois, B. Drevet, R. Einhaus, and Y. Delannoy, 2012. Dopant specifications for p-type UMG silicon:mono-c vs multi-c, *Proceedings of the 27th EUPVSEC*, Frankfurt, Germany, pp. 1002–1005.
106. J. Kraiem, P. Papet, J. Degoulange, M. Forster, O. Nichiporuk, D. Grosset-Bourbange, F. Cocco, and R. Einhaus, 2012. World class solar cell efficiency on n-type CZ-UMG silicon wafers by heterojunction technology, in *Proceedings of the 27th EUPVSEC*, Frankfurt, Germany, pp. 657–660.
107. M. Forster, J. Degoulange, R. Einhaus, J. Veirman, N. Enjalbert, and R. Cabal, 2013. P-type and N-type CZ solar cells made with 100% Photosil silicon: Impact of boron concentration, in *Proceedings of the 28th EUPVSEC*, Paris, France, pp. 1431–1434.
108. R. Einhaus, J. Kraiem, B. Drevet, F. Cocco, N. Enjalbert, S. Dubois, D. Camel, D. Grosset-Bourbange, D. Pelletier, and T. Margaria, 2010. Purifying UMG silicon at the French PHOTOSIL project, *Photovolt. Int.*, 9, 60–65.
109. T. Margaria, F. Cocco, J. Kraiem, J. Degoulange, D. Pelletier, D. Sarti, Y. Delannoy, C. Trassy, L. Neulat, and R. Einhaus, 2010. Status of the Photosil project for the production of solar grade silicon from metallurgical silicon, in *Proceedings of the 25th EUPVSEC/5th WCPEC*, Valencia, Spain, pp. 1506–1509.
110. A. Turenne, S. Nichol, and D. Smith, 2012. Method of purifying silicon utilizing cascading process, US Patent 8,580,218.
111. T. Yoshikawa and K. Morita, 2003. Solid solubilities and thermodynamic properties of aluminum in solid silicon, *J. Electrochem. Soc.*, 150, G468–G471.
112. T. Bartel, K. Lauer, M. Heuer, M. Kaes, M. Walerysiak, F. Gibaja, J. Lich, J. Bauer, and F. Kirscht, 2012. The effect of Al and Fe doping on solar cells made from compensated silicon, *Energy Procedia*, 27, 45–52.

4 Elkem Solar and the Norwegian PV Industry through 40 Years (1975–2015)

Bruno Ceccaroli and Ragnar Tronstad

CONTENTS

4.1   INTRODUCTION

This chapter deals with 40 years of history (1975–2015) and the remarkable technical and industrial achievement of a Norwegian company. Few countries besides the industrial superpowers, for example, the United States, China, Japan, and Germany, have made a more significant contribution to the solar industry than Norway. What are the reasons for a country of less than five million inhabitants, most of the year living in places where there is less sunlight, to play such a role in the history of photovoltaics (PV)? This seems an unlikely history to most outside observers, who may hardly imagine the country as a sunny place. The background to this remarkable and astonishing contribution is a century-old metal industry, built to make use of the country's huge hydropower resources. At the beginning of the history, there is a project striving to invent a process aiming for the production of silicon that is pure and cheap enough for the mass production of solar cells. At the end of the history (concomitant with the edition of this book) is a world-leading integrated solar company and a myriad of other companies, producing everything from silicon to solar modules and systems.

In 1975, the OPEC embargo made the Western world acknowledge its dangerous dependence on oil and triggered several initiatives to develop alternative methods to produce energy, including solar electricity (PV). Growing concern over environmental issues and climate change has since then steadily strengthened the need to accelerate the energy transition, conferring to solar energy a central role.

For those aspiring to a revolutionary silicon process for solar cells in 1975, it was natural to approach the company Elkem, at the time one of the most accomplished and recognized Norwegian technology companies. Founded in 1904 to develop technology solutions to make use of the huge national potential of hydropower, its first success was the Soederberg electrode. The company sold its concept all over the world through various license arrangements, concerning homemade carbon paste and submerged arc furnaces. More recently, the company had evolved from a technology provider to a new position as an industrial manufacturer for metals and ferroalloys, including silicon.

The technical development at Elkem on the way to solar-grade silicon (SoG-Si) at first (1975–1985) was very rapid, generating essential results that paved the road to the current process (2015). Then, for a long decade (1986–1999), doubt and hesitation prevailed before the company resumed development work at full speed (2000–present). At the end of the first period (1985), the company had already worked out a downstream integration strategy. However, the reader will have to wait until the end of this history (2015) to see it fully accomplished under the umbrella and control of one owner and producing everything from silicon to modules through ingots, wafers, and cells. Meanwhile, the reader will also learn about the rise and fall of younger and ambitious enterprises challenging and shadowing the old company, which had fostered them. Therefore, this chapter includes other Norwegian industrial attempts to get a strong foothold in the solar industry, particularly the spectacular emergence of Renewable Energy Corporation (REC), including its predecessor and successors (1994–2015). Even so, the main topic of this chapter remains the history of Elkem Solar Silicon™ (ESS™), the invention of its process, its development, and its upscaling to the current industrial stage. Other chapters and books devoted to REC's story and other Norwegian contributions to the solar business and science deserve to be written, hopefully sooner rather than later (see Box 4.2 and Chapter 5 for fluidized bed reactor (FBR): development by REC).

4.2   METHODOLOGY

Both authors of this chapter have been deeply involved in the history described herein. Together, they cover most of the 40 years described in the chapter, that is, R. Tronstad, 1980–1985 and 2000–2015 at Elkem, and B. Ceccaroli, 1986–1999 at Elkem and 2000–2007 at REC. A nonnegligible part of the text builds on their personal experience and memory, bringing to the description on the one hand the benefit of highlighting the decision processes within the company in an array of conflicting interests, but on the other hand adding the risk of subjective and selective memory. Being aware of this potential risk, the authors have taken great care of recollecting and cross-checking their own memory with the available data and records filed with the company. They consulted more than 500 documents, including scientific and technical reports, memorandums and notes, calls to meetings and minutes of meetings, and letters from and to external parties. All the files are registered and well preserved at the central archive of Elkem Research/Elkem Technology in Kristiansand. Most of the consulted documents are not public literature. When necessary for the sake of historical accuracy, the authors may refer to these files as "internal" documents with a clear identification code although the file might not be accessible to all readers. The histories of Crystalox Ltd, REC, and subsidiaries (see Boxes 4.1 and 4.2) were established in close cooperation with key employees and managers at those companies. These contributions are referred to as "private communication." The authors have tried to report the historical facts as faithfully as possible. However, the authors are solely responsible for the appreciation and interpretation of the events as they are described in this chapter.

4.3 DOW CORNING (1975–1980)

4.3.1 INDUSTRIAL CONTEXT OF ELKEM: DOW CORNING

Since its foundation in 1904, under the name Det Norske Aktieselskab for Elektrokjemisk Industri or Elektrokjemisk AS, Elkem had been a technology-driven company. The development of continuous electrodes for submerged arc furnaces—named the "Soederberg electrode" after its inventor—was the first invention of historical value and the basis for taking Elkem to the position as global leader in metallurgical engineering. In the early days of electric furnace operations, electrodes were used one by one and when an electrode was worn out, the furnace had to cool down before a new electrode could be installed, a procedure that could take several days. The new technology made continuous electrode operation possible. The Soederberg technology was developed for, among others, ferroalloys and aluminum processes and brought considerable cost savings compared to the former, stationary, electrode system. Over time, furnaces for the production of pig iron, ferromanganese, silicomanganese, ferronickel, ferrochromium, ferrosilicon, etc. with Soederberg technology spread through all parts of the world. Most plants were erected at remote places, close to the ores and/or power sources. The large variety of both raw material qualities and climatic conditions posed numerous challenges to the engineering company, but at the same time offered great opportunities to acquire a broad and unique experience. Using industrial data collected from operations at these plants, Elkem could build up reliable algorithms for the design of new furnaces. Moreover, engineering data could be obtained by pilot-scale experiments as testing of less known or unknown raw materials in Elkem's own pilot plant became a standard and important procedure for correct industrial design. In 1972, Elkem had merged with Christiania Spigerverket, a mining and iron company and as such a user of Elkem's furnace technology. With this strategic move, the new company Elkem Spigerverket added a significant mining and manufacturing capacity within steel, pig iron, ferroalloys, and silicon metal to its excellence in engineering [1].

Elkem's global experience was well known in the industry and was the reason that Dow Corning in the 1970s considered Elkem as a cooperation partner for the development of a new process concept for SoG-Si. Dow Corning, a North American 50%–50% joint venture between Dow Chemical and Corning Glass, was, and still is, a worldwide leader for the manufacturing of polysiloxanes (commonly designated as silicones) and semiconductor-grade silicon (commonly designated as polysilicon) and, as such, was also the largest single consumer of chemical-grade silicon, a variant of metallurgical-grade silicon (MG-Si).

As a response to the unexpected OPEC oil embargo and price dictate, the U.S. Department of Energy (DoE) had launched a low-cost photovoltaic conversion program for terrestrial electricity generation. In 1974, the cost level for solar cells in space applications, then the leading photovoltaic application, was $74/W. The DoE Low Cost Silicon Solar Array (LCSSA) project, directed by the Jet Propulsion Laboratory (JPL), targeted 500 MW of annual photovoltaic power installations at a solar array cost of less than $0.5/W by 1986. Solar cells for space applications were made of single-crystal silicon available from the solid-state electronic industry. Silicon solar cells and transistors have a common silicon feedstock, that is, polysilicon, an ultrapure grade achieved by *chemical vapor deposition* (CVD) of purified chlorosilanes in a bell-jar reactor. Commonly designated as the Siemens process (named after the company, which first contributed to the development of the bell-jar reactor), this process was considered both expensive and highly energy consuming [2–4]. To reach the goal assigned to the LCSSA by the DoE, the price of SoG-Si had to be reduced from the average market price of $65/kg to less than $10/kg. This cost reduction was not possible with the existing Siemens process. Among the contributors to the LCSSA, Dow Corning proposed to explore whether pure raw materials applied to the direct process for MG-Si could achieve this goal and replace the Siemens process for SoG-Si (see Chapter 5 on the polysilicon processes' history and status).

4.3.2   FROM IDEA TO REALITY

The process idea contemplated by Dow Corning as indicated in Figure 4.1 included three steps (see technical details in Chapter 2) [5]:

- The first step was the selection and purification of the raw materials, that is, quartz as the Si source and carbon as reductant to the oxide.
- The second step was a close to a standard submerged electric arc furnace process; raw materials are charged to it and react at high temperature to form a crude silicon alloy.
- The third step was imagined as a directional solidification (DS) to further refine the crude silicon as it was believed that the less segregating elements were low enough in the raw materials.

The cooperation between Dow Corning and Elkem (Elkem Spigerverket) started in 1975 to develop the "direct arc reduction" (DAR) as the next generation of SoG-Si process.

Dow Corning had limited experience in silicon production by the metallurgical route. They needed external support. Therefore, in February 1975, they presented their idea at a meeting with Elkem, officially as a part of a tender procedure for a broad supply package [6], including

- A 50 kVA bench-scale furnace
- A 5–600 kVA pilot furnace
- A test run program at the Elkem R&D center (Kristiansand, Norway)
- A 15–20,000 metric tons per year (MT/year) furnace for high-purity Si

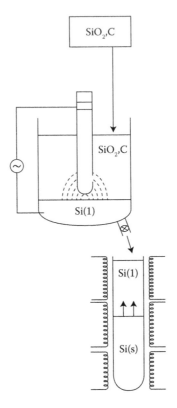

FIGURE 4.1 Schematic view of Dow Corning's contemplated low-cost SoG-Si process in 1975. (Adapted from V. D. Dosaj, L. P. Hunt, and A. Schei, 1978. High-purity silicon for solar cell applications, *JOM*, 30, 8–13.)

Two months later, Dow Corning returned to Elkem for further discussion. On the second day of their visit, Dow Corning's representatives had the opportunity to inspect the test facilities at Elkem R&D in Kristiansand, Norway, equipped with electrical furnaces ranging from 50 to 1500 kVA. As a result of the discussions with Elkem's R&D staff, the 50 kVA furnace held most of Dow Corning's attention. Elkem had never offered or delivered test furnaces to customers as running test facilities was regarded as an integral part of the commercial package when selling furnaces. Elkem accepted the adjustment of its business model. After the 3-day visit, Elkem agreed with Dow Corning on a tentative program for delivering bench-, pilot-, and commercial-scale furnaces with the following schedule:

- 50 kVA furnace installation in 1977 [6]
- 500 kVA pilot furnace in 1979 [7]
- 4–5 MVA commercial furnace in 1984 [8]

The bench scale was later increased from 50 to 150 kVA [9,10].

In the early 1970s, Elkem had limited experience with silicon refining except for the Silgrain® leaching process developed by Spigerverket at the Bremanger plant

[11,12]. Dow Corning had some minor experience with chlorine refining (impacting less noble elements than Si, e.g., Al and Ca) and some directional solidification (DS) know-how (impacting more noble elements than Si, that is, transition elements and most of metals). The cooperating parties were aware of the merit of these purification techniques. None would be efficient for boron removal; therefore, the raw material quality had to take care of this critical element (acceptor, dopant) whose concentration in silicon determines semiconductor performances [13].

The first bench-scale tests took place in March–April 1976 at the Elkem R&D center. A mix of pure quartz (optical quality) and purified charcoal was smelted in a 50 kVA furnace. A severe accumulation of the solids at the bottom of the furnace and unreacted quartz on the sidewalls of the crater made proper operational conditions impossible [14]. The selected purified charcoal showed, in standard test setup, a significantly reduced SiO reactivity as compared to unpurified charcoal.

Moreover, the purification process applied to the charcoal caused a serious reduction of its mechanical strength [15]. Consequently, new test plans had to include an optimization of the purification temperature for charcoal. However, further bench-scale testing of purified charcoal did not bring conclusive improvements to the furnace operation. In these experiments, the used charcoal had been purified at temperatures in the range between 1700°C and 2500°C. The bad performance of purified charcoal probably masked the effect of other variables such as furnace operation strategy, electrode/furnace shell diameter, taphole design, and lining quality [16,17].

From the small amounts of silicon produced during the smelting campaigns, samples could however be analyzed. Analyses showed promising results with lower B and P content than standard MG-Si [18]. Some of these silicon amounts were submitted to DS in an attempt to achieve a higher purity and then be crystallized for wafering. Wafers from the resulting ingot were analyzed to contain less than 1 ppba of transition metals, but exhibited high values for B (9 ppma) and P (12 ppma). Solar cells made of these wafers showed an absolute reduction of energy conversion efficiency of 2% as compared to solar cells made with standard commercial feedstock (efficiency measured under standard conditions of 11% instead of 13%) [19].

Installation of the new 150 kVA furnace at Dow Corning was completed in January 1978 [20] and further test work could be conducted in the United States. Significant efforts were put on lining [21] and on furnace design in order to receive appropriate pelletized raw material charges. During the furnace construction, a survey conducted on more than 90 different silica sources revealed that several sources could meet the purity goals. Two commercial sources of quartz were then retained as meeting all the requirements on cost, purity, and chemical reactivity. As carbon sources, activated carbon, carbon black, wood, sugar charcoal, lignite, and petroleum coke were examined, but only activated carbon and carbon black showed acceptable purity and reactivity. Both qualities of these reductants were tested as pellets containing sucrose as binder [22]. Replacement of purified charcoal, from the initial bench-scale tests, with these carbon black–sucrose pellets improved furnace operation. Tests in the 150 kVA furnace demonstrated continuous and controllable processing for 60 hours at a power consumption of 26 kWh/kg Si, an improvement of 10–15 kWh compared to the purified charcoal campaigns. More than 100 kg Si could be produced per run with impurity content between 50 and 100 ppmw for Fe and Al and less than 10 ppmw for other elements. Further refining in a Czochralski (CZ) puller reduced the impurity level of most elements to an acceptable level for solar cells. However, Al (1 ppma), B (7 ppma), and P (0.5 ppma) were still judged too high for that purpose.

After a second pass through the CZ puller, a single crystal could be achieved. The ingot was sliced into wafers, from which cells were produced and tested. Cell performances at par with those of commercial cells demonstrated the potential of the metallurgical route. Depending on cell technology and the position of the wafer in the ingot, cell efficiencies ranging from 8.2% to 14% were measured before the project was closed.

4.3.3 DAR Project Completion

In the final report [22] to DoE/JPL, Dow Corning experts J. P. Hunt and V. D. Dosaj stated that the objectives of the LCSSA project had been achieved, demonstrating the technical feasibility of a metallurgical process for high-volume production of SoG-Si.

In 1979, after the completion of the DAR project with Elkem, Dow Corning decided without explicit reason to withdraw from further development of the metallurgical SoG-Si route. Looking back at the work done by Dow Corning and Elkem, it is obvious that the results clearly demonstrated the potential of the metallurgical refining route as a substitute to the Siemens technology and paved the road to many successors aspiring to similar goals. However, it would take another 30 years before any commercial plant for metallurgical SoG-Si processing would start up.

4.4 COOPERATION WITH EXXON: FURTHER CONTRIBUTION TO THE JET PROPULSION LABORATORY (DoE) (1979–1985)

4.4.1 Project Initiation and Context

In the meantime, A/S Norsk Esso (a Norwegian subsidiary of the North American petroleum giant Exxon) had contacted Elkem. As a condition for exploration and drilling rights in the Norwegian offshore fields and territories in the North Sea, the oil companies had to support land-based projects benefitting the development of new industry in Norway (see also Section 4.5.1 about Norsk AGIP/Eurosolare). Solar Power Corporation (SPC) was another subsidiary of Exxon. Dedicated to solar power, SPC had been part of the photovoltaic industry since 1970, establishing sales offices in 35 countries and following the cooperation between Dow Corning and Elkem on the DAR process with great interest. In September 1979, at a meeting in Oslo, Exxon proposed a 2-year development program between Exxon, Dow Corning, and Elkem [23]. Elkem's proposed role should be to further develop the furnace technology and bring production expertise, Dow Corning should focus on raw material purification and handling, and Exxon should take care of crystal growing, solar cell processing, and project management. Dow Corning declined the invitation to take part in the new project as the management had decided to withdraw from all metallurgical solar-grade refining activity. However, Dow Corning did not object to an independent continuation of the project on the DAR process by other parties. As the former Dow Corning/Elkem project had been financed under the DoE/JPL/LCSSA program, the results were open and free to use by everyone. Therefore, the two other parties could, with Dow Corning's consent, enter into a bilateral cooperation agreement and pursue the new proposal.

At a meeting in February 1980 a steering committee and the project management were appointed. The project leaders were Anders Schei/Elkem and John Dismukes/Exxon [23–25].

An updated project proposal issued by Exxon in January 1980 [26] included six tasks:

1. Raw material selection and pretreatment—joint program
2. Arc furnace operation—Elkem
3. Posttreatment—joint program
4. Crystal growth—Exxon
5. Solar cells—Exxon
6. Engineering and economic analysis—joint program

The chain of process steps presented by Exxon followed the track explored by the Dow Corning–Elkem project and endorsed more or less the same financial goals. Some new or additional purification alternatives were, however, suggested as shown in Figure 4.2.

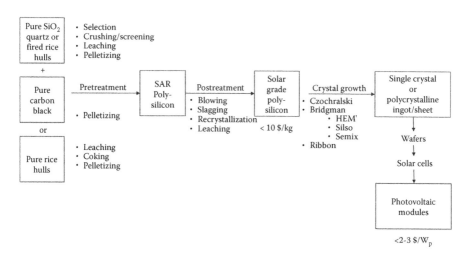

FIGURE 4.2 Exxon's proposed process for the production of low-cost photovoltaic modules—cost in 1980 dollars. SAR, submerged arc reduction. HEM, Silso, and Semix were the existing commercial options for Bridgman techniques. (Adapted from A. Schei, 1980. Samarbeid Exxon-ES, *Memo, 08.05.1980.*)

In May 1980, John Dismukes paid a visit to Elkem. No formal contracts or research agreements had been signed yet. His visit triggered a new discussion within Elkem on the relevance of this follow-up project now that Dow Corning had adjourned its participation and officially had withdrawn definitively from the business. The internal discussion addressed three main concerns [28]:

- Would a relatively modest company like Elkem be able to establish a competitive position in a market where large and powerful companies could spend significantly more important resources in R&D and capital? Could Elkem be an equal partner to Exxon?
- Why had Dow Corning withdrawn from the business if the future of low-cost SoG-Si looked so bright?
- What was the real motivation of Exxon? A real interest in the DAR/SAR technology or just the drilling rights in the North Sea? (DAR was renamed to SAR—submerged arc reduction—in the Exxon project.)

Through several technology inventions, Elkem had gained a unique position as compared to other MG-Si producers. In the 1970s, the Elkem "split furnace body" design [29] had been developed by Harald Krogsrud independently of the ongoing solar projects. In the same period, the Silgrain plant for hydrometallurgical purification of FeSi had been installed at the Bremanger plant [11,12], and the test facilities in Kristiansand (R&D center) had for some years been utilized for detailed studies of the MG-Si and SoG-Si process.

Furthermore, the cooperation with Exxon took place in a period when Elkem through mergers and acquisitions became a global giant within ferroalloy technology and manufacturing. We have already mentioned the merger between Elkem and Christiania Spigerverket in 1972. In 1981, Union Carbide's ferroalloy division and its plants in the United States, Canada, and Norway producing ferroalloys and MG-Si were added to the Elkem group. This expansion strategy called for a stronger position in the more advanced silicon market. To take the lead in developing a new metallurgical route for SoG-Si was therefore an attractive opportunity although it could be expensive and would need verification in testing solar cells, which was far beyond Elkem's technical capability. Therefore, a joint project with Exxon seemed a good fit as both parties obviously had complementary interests.

Market development of solar cells sounded appealing for an eager silicon company as no serious competitor to crystalline silicon cells seemed capable of entering the market [30]. That Dow Corning had stopped the activities on SoG-Si was perceived more as a strategic commercial decision than a technical failure of the joint project. Dow Corning's core businesses were silicones and polysilicon for semiconductors, for which Dow Corning was without contest a market leader. No doubt the company intended to preserve this strong position and might feel not financially able to embrace a more widespread and uncertain business in the energy sector.

All these considerations may explain why Elkem's management found worthwhile continuing the development on the SoG-Si process in cooperation with Exxon.

4.4.2   ARC-FURNACE PROCESS DEVELOPMENT

The Exxon–Elkem project was officially launched with the signing of an R&D agreement in February 1981. However, a common understanding had been already reached between the parties in June 1980, allowing some initial work in both Norway and the United States. Bench-scale furnace tests had been run since August 1980 with SAR 1 (test number in the Exxon/Elkem program) to SAR 8. In the first eight tests, quartzite was added as lump and two different carbon black qualities were added as sugar-bounded pellets. No breakthrough results were observed until SAR 9 in which combined quartzite–carbon pellets replaced the carbon black pellets. Sugar was still used as binder and lumpy quartzite was used to adjust the stoichiometric balance in the charge. SAR 9 was very successful and set a new standard for the following test programs [31]. In 1980 and the first part of 1981, little focus was directed toward pure furnace operation; the same laboratory-scale equipment as prior to the Exxon project was used. The furnace was equipped with a rotating top-ring to force the charge sinking in the furnace without the use of stoking equipment. The industrial stokers were normally heavy steel arms forcing the pre-reduced charge downward into the hot reactive inner part of the furnace. Iron is one of the critical elements in solar cells, and by avoiding the use of stokers, the probability of reaching the Fe specification increased. However, most tests showed difficulties after 1 day of operation with buildups of crusts around the electrode that occasionally broke off and accumulated on the top of the charge. This made the carbon balance difficult to control and hydraulic stokers were reinstalled.

SAR 1–14 had been run in the small laboratory furnace. In the late summer of 1981, a new 150 kVA (equivalent to 100 kW) furnace was installed and made ready for high-purity Si production. The furnace was placed in a clean area where operators had to wear special clothes and only trained personnel were admitted. The furnace lining in contact with the charge and the melt was selected among purified materials to satisfy the high-purity requirement. SAR 15 in the new equipment and also the tests that followed did not show the expected improvements. The main reason seemed related to technical problems. However, it also turned out to be a lack of understanding of which parameters influenced the performance in high-purity silicon production. A new series of 16 tests conducted during the autumn of 1982 gave a clearer feedback. The most important results were the following [32]:

- Balance the charging flowrate with the power input
- Control the charge sinking with traditional mechanical stoking instead of the split body system (for small furnace systems)
- Control the furnace electrical resistance by introducing materials, which can stabilize the arc

Applying these recommendations to the new 100 kW/150 kV furnace, the performance gradually improved, and the real breakthrough arose at SAR 30 with the introduction of purified lime in the charge. The next five tests were run with equal charge composition, and except SAR 31, all tests showed stable performance with acceptable power consumption (corresponding to 12–14 MWh/MT Si in industrial-sized furnace) [33].

4.4.3 INFLUENCE OF RAW MATERIALS AND OTHER INPUTS TO THE FURNACE

Raw materials included quartz, pure rice hulls, fired rice hulls (all three SiO_2 sources), carbon black (reductant), purified lime (additive input), purified electrodes (power input), graphite construction materials, and lining materials (furnace components).

Two North American quartz sources were used in most of the smelting tests, that is, Mount Rose quartz from British Columbia (Canada) and Malvern quartz from Arkansas, USA. Both qualities had B and P content of approximately 1 ppmw. These qualities were used as lumps for adjusting the stoichiometric balance. The main silica source was however pellets of milled quartz. In most of the tests, Iota quartz sand of optical quality was used. High-quality Norwegian sources from Bryggja and Drag were upgraded by flotation during a ton-scale experiment at SINTEF (a Norwegian contract research organization associated with the technology university). The intention was to duplicate the quality of Iota quartz, which had resulted in low B content in the SAR experiments. No purification with respect to B and P was obtained by flotation. However, with initial levels of 1–2 ppmw for B and less than 1 ppmw for P in Drag quartz and less than 1 ppmw for both B and P in Bryggja quartz, it was decided to test these Norwegian sources in SAR experiments as B content in Iota quartz was only slightly lower, that is, 0.9 ppmw [33,34]. The Norwegian quartz qualities were tested with SAR 34. Except for some iron pickup during smelting, the furnace operation was comparable to Iota quartz and B and P content in leached Si (SILSAR) were approximately 1 ppmw for both elements [35].

4.4.4 SILICON REFINING

Elkem's responsibilities in the cooperation with Exxon were not only related to furnace operation but also to the necessary refining steps to reach the specification for SoG-Si. The initial impurity levels defined by Exxon had an upper level of 100 ppmw for each iron, aluminum, carbon, and the sum of all other elements. For the doping elements, 1 ppmw was defined for both boron and phosphorus [36]. It was obvious that these limits could not be obtained out of the furnace. Therefore, additional process steps for refining were suggested as shown in Figure 4.3. Common for the two alternative processes is the Silgrain pure Si crystals step, meaning that impurity-rich grain boundaries between Si grains may be digested by leaching liquors. As metallic impurities move to the grain boundaries during solidification, the leaching process is able to considerably reduce the impurity level. However, the potential efficiency of the method depends on critical parameters such as the composition, the thickness, and the distribution of the grain boundaries. In alternative 1 and 2, calcium oxide is added to the melt, with the addition of lime to the furnace charge pursuing two purposes, that is, stabilization of the electric arc and making the cast silicon leachable. A more detailed description of the leaching process is given in References 37 and 38 and in Chapter 3.

Alternative 2 introduces a different smelting process, suggesting an aluminum-rich SiAl40 alloy. Experience had shown that the SiO reactivity requirement for carbon in producing SiAl40 is much lower than for pure silicon. This option could

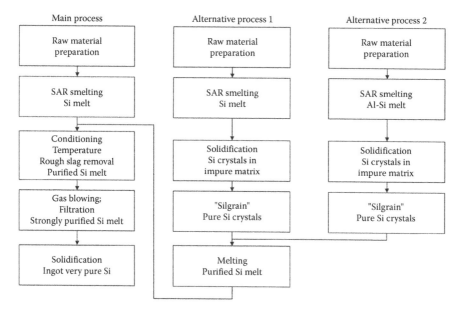

Main process	Alternative process 1	Alternative process 2
Raw material preparation	Raw material preparation	Raw material preparation
SAR smelting Si melt	SAR smelting Si melt	SAR smelting Al-Si melt
Conditioning Temperature Rough slag removal Purified Si melt	Solidification Si crystals in impure matrix	Solidification Si crystals in impure matrix
Gas blowing; Filtration Strongly purified Si melt	"Silgrain" Pure Si crystals	"Silgrain" Pure Si crystals
Solidification Ingot very pure Si	Melting Purified Si melt	

FIGURE 4.3   Three alternative processes for smelting and purification of MG-Si to SoG-Si suggested in the Exxon proposal of 1980. (Adapted from A. Schei, 1983. Exxon–Elkem joint R&D program on low-cost silicon for solar cells, *Elkem Report No. 10 to Steering Committee/ Internal Report F79/83a.*)

open for petroleum coke a quite pure reductant but which is not easy to use in large quantities in silicon production, because of its low SiO reactivity. Following this option, silicon and aluminum would then have to be separated from each other, for instance, during solidification. If 25% of the silicon could be precipitated and separated as pure Si crystals, the remaining alloy would be sold to the aluminum industry. To verify the distribution of impurities between crystal and melt, the Norwegian contract research foundation SINTEF was engaged. The distribution coefficients were determined experimentally at 1020°C, which according to the binary phase diagram should give an equilibrium crystal content of 24%. The study predicted that a SiAl40 alloy containing 1% transition elements as impurities would result in 100 ppmw impurity in silicon crystals. Aluminum, calcium, and carbon would however be higher and would need separate purification steps. Boron content would also be an issue requesting additional specific treatments [31]. No further studies were however engaged to validate this method (alternative 2).

Removal of carbon was a difficult task. Owing to the massive presence of carbon in the smelting furnace, 100–2000 ppmw would always be present in the liquid melt [39]. Carbon concentration at saturation is temperature dependent; high temperature increases the solubility, and solubility in the solid phase is approximately 5 ppmw compared to approximately 50 ppmw in the liquid phase at the solidification temperature. Operation with saturated solutions will therefore always lead to precipitation of solid SiC particles. The presence of SiC particles is detrimental to the wafer production and the solar cell performance causing ultimately short circuits in the solar cells. The methods envisaged for carbon removal included the following:

- Filtration
- Gas blowing and slagging
- Oxidation
- Directional solidification (DS)

DS was the only method with high C purification capability, and less than 0.1 ppma C (as SiC) was detected after treatment. The amount of dissolved carbon was 10 ppma. The other methods of carbon removal were at that time not experimentally sufficiently proven to be efficient [39].

With boron content of approximately 1 ppmw in quartz and minor amounts in other raw materials, it was expected to reach 2 ppmw in Si tapped from the furnace. DS experiments [34] had been run by Exxon where the influence on B and P content on solar cell efficiency showed a decreasing efficiency from approximately 12% to 11% when B content increased from 1 to 2 ppmw. Further refining experiments were therefore initiated to decrease B by means of two process options:

- Gas injection of H_2–H_2O containing gas
- Slag treatment with Na_2O–SiO_2 addition

The gas treatment was performed in accordance with a published method [40] but did show an effect only at high B concentrations in Si. Even if there was a reduction in B content during the slag treatment from 2 to 1.7 ppmw, the results were not conclusive as the B level was outside the analytical capability at Elkem (see Table 4.4). Further studies on the slag treatment were suggested, but soon postponed to later programs.

4.4.5 CHEMICAL ANALYSIS AND CHARACTERIZATION

Moving from the production of MG-Si with an impurity content of 1%–2% to a solar-grade quality considerably influenced the furnace operation on the selection of raw materials, consumables, equipment, process control, and the verification of product quality. Analytical methods in use at Elkem were directed toward the common elements present as impurities in standard MG-Si grades with few elements specified below 0.1%. For solar grades, the upper limit on most elements was in the range 100 ppmw, and 1 and less than 1 ppmw for B and P, respectively. One of the most important control parameters for SoG-Si is chemical analysis of the doping elements (principally B and P). Elkem at that time had no analytical methods for B and P in the sub-ppm level. During the first SAR experiments, two quartz qualities were tested and samples were analyzed by two different specialized laboratories, one recruited by Exxon and one by Elkem. The differences in treating and analyzing high-purity samples are illustrated by the results shown in Table 4.1.

With the present methods, both the time needed and the accuracy on P and B analyses were inappropriate for process control. In determining the B and P

TABLE 4.1
Analyses of Two Quartz Qualities
Performed at Two Different Laboratories

Element	Quartz Quality A		Quartz Quality B	
(ppmw)	Elkem	Exxon	Elkem	Exxon
B	8.5	0.5	7	0.5
P	14.0	<1	6	<1

Source: Adapted from J. P. Dismukes, 1983. Final Exxon report to the steering committee for the quarter October 1, 1982 to December 31, 1982.

concentrations for SAR-Si (from arc furnace) and SILSAR (leached SAR-Si), the calculation proceeded backward from the Hall measurements on wafers on given points of the second recrystallized ingot [41]. Samples were sent to Exxon and evaluation of the tests could not be finished until analyses were returned weeks later.

In attempts to develop chemical analyses, samples from Elkem were also analyzed by different techniques and by different institutes. In addition to the Exxon laboratory, spark source mass spectroscopy was performed at IFE (Institute of Energy Technology, also a Norwegian institute executing research contracts), atomic absorption spectroscopy at SINTEF, colorimetric determination of P at Fiskaa Verk (Elkem plant), and x-ray fluorescence spectroscopy at the Elkem R&D center.

In a standard MG-Si production, the reactivity of coke and coal is a critical parameter for obtaining efficient furnace operation with high yield and low power consumption. The term "reactivity" is in this context referred to as the reaction between SiO and C to form SiC:

$$SiO + 2C \rightleftharpoons SiC + CO \tag{4.1}$$

In early days, the reaction between CO_2 and C to CO was rather used:

$$CO_2 + C \rightleftharpoons 2CO \tag{4.2}$$

under the assumption that there was a correlation between the "CO_2 reactivity" and the "SiO reactivity." This assumption was however purely empiric. In 1976, J. K. Tuseth and O. Raaness published a method for direct measurement of "SiO reactivity" [42]. In the Exxon–Elkem cooperation, this latter method has been used for all reactivity measurements of reduction materials.

4.4.6 VERIFICATION OF Si QUALITY IN SOLAR CELLS

Exxon had worked out a specification for SoG-Si, but this was to a large degree based on the expected capability of metallurgical processing. According to this specification, several hundred ppm of impurities were acceptable. Davis et al. had already in 1980 [43] presented a study on how impurities could influence solar cell performance. Specifications resulting from this work were expressed at the ppb level. Therefore, a final verification of Si quality by the production and the physical evaluation of standard solar cells were necessary.

The best Si qualities were produced in the last tests, and leached SILSAR products from SAR 29, SAR 33, SAR 34, and SAR 35 were treated in Bridgman crystallizers to generate wafers for 2×2 cm^2 solar cells. Some samples were crystallized twice, and all results were compared with a reference cell made of "space-quality" silicon by Applied Solar Energy Corporation (ASEC). ASEC was a principal evaluator of silicon materials for the JPL-monitored LCSSA project.

Figure 4.4 is the picture of a module made from cells manufactured with Exxon–Elkem silicon.

As seen from Table 4.2, the quality of solar cells made from the produced SILSAR is only 1% efficiency away from the monocrystalline cells based on the best semiconductor Si quality. The table also shows the importance of repeated Bridgman crystallization. Solar cells made from SILSAR 34 obtained an improvement of 3% after a second Bridgman crystallization.

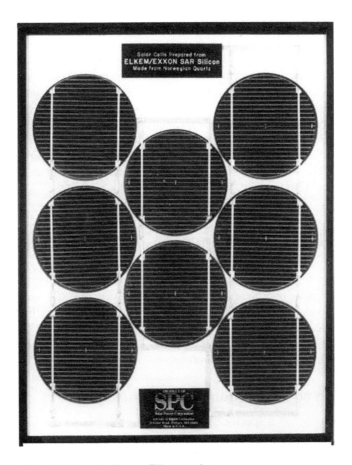

FIGURE 4.4   Module from the Exxon–Elkem project.

TABLE 4.2

Solar Cell Data from Tests SAR 29, 34, and 35 Compared with Monocrystalline Reference of "Space Quality" Silicon by Applied Solar Energy Corporation

Ingot Type	Resistivity p-Type (ohm-cm)	Short Circuit Current (mA/cm²)	Open Circuit Voltage (mV)	Fill Factor	Cell Efficiency (η) (%)
SILSAR[a]-34 1× Bridgman	0.25	27.7	587	0.72	11.7
SILSAR-29 2× Bridgman	0.5	30.9	597	0.78	14.3
SILSAR-34 2 × Bridgman	0.3	31.3	610	0.78	14.7
SILSAR-35 2× Bridgman	2.4				13.9
ASEC reference (Czochralski)	1–2	33	590	0.8	15.7

Source: Adapted from J. P. Dismukes, 1984. Solar cell performance assessments of Elkem–Exxon silicon, *Internal Report F114/1984.*
[a] SILSAR means leached SAR-Si.

4.4.7   INDUSTRIAL PLANT

The final objective for Elkem in the cooperation with Exxon was related to industrialization. Bremanger was at an early stage, and looked upon as the most suited place for the new plant, and this was mainly because of their background in hydrometallurgical refining. During the second quarter of 1982, a preliminary cost estimate was made. The total cost for an SoG-Si plant with a capacity of 1000 MT/year was estimated at 80 million NOK, including ground work, building, and installation [45]. When the Exxon–Elkem project ended in December 1983, the market was still too low to initiate industrialization of specialized SoG-Si materials. The plant in Bremanger was therefore never built. In 1984, Exxon sold its assets in SPC to Solarex and decided the same year to stop all photovoltaic activities.

The SAR campaigns, combined with an efficient refining process, had demonstrated the possibility of using silica–carbon pellets in the submerged arc furnace for the production of pure silicon for solar cell applications.

According to the original project plans, Elkem should have taken the project to the next gate and established engineering data based on pilot-scale experiments in a 1500 kVA furnace. This experiment never started as part of the Exxon–Elkem cooperation, but was carried out by Elkem alone in 1986. Both Bryggja and Iota quartz were considered for this test, but due to a price level of 30,000 NOK/MT, the test were carried out with a lower-grade quartz from Norfloat, Norway [46,47]. Pellets of quartz, coal, and binder led to good results in the 1500 kVA furnace with no buildup of SiC in the furnace. Elkem decided therefore to build an industrial-sized pelletizing plant at Fiskaa and rebuild an old plant for silica pellets into a modern plant for raw material pellets. Production of quartz/coal pellets followed the same ideas as tested in the pilot plant, that is, use of sugar as a clean binder, and obtained pellets of similar quality. Furnace Nr. 10 at the Fiskaa plant, a three-phase furnace of 20 MW, was selected for the test. It turned out to be difficult to use a large amount of pellets in the charge. When the pellet temperature reached 700–800°C, sugar decomposed and lot of fines build up in the furnace, which resulted in a violent fire. However, in the first part of the test, the operation gave the best power consumption figures obtained for Si production in this furnace. Implementing the learning from this test reduced power consumption from typically 14–12 MWh/MT Si also for standard charge composition.

From an SoG-Si production point of view, the poor pellet behavior stopped further development on direct furnace production of high-purity Si based on pure raw materials. In later projects, Elkem would rather utilize post-taphole processes for purification of MG-Si to high-quality SoG-Si.

4.5 NORSK AGIP/EUROSOLARE AND TEXAS INSTRUMENTS AND A NEW SLOW DOWN (1991–1996)

4.5.1 NORSK AGIP/EUROSOLARE

A period without PV activity at Elkem followed from 1986 until the end of the decade. In fact, it was during this period politically incorrect in the higher echelons of corporate management to mention solar energy or SoG-Si. Therefore, a few courageous or visionary technologists who still wanted to keep a minimum of activity in this area, mainly in technology survey, had to hide their activity behind more popular technology topics. Some of the ceramics and advanced materials produced in minute quantities mainly in Japan and the United States requested pure silicon as raw material. Targeting these potential applications, the technologists therefore saw a chance to keep dealing with their passion for pure silicon hoping that solar energy would sooner or later make a comeback. This was a wise attitude proving to be invaluably helpful two decades later.

After this long period of inactivity, a project with Norsk AGIP and Eurosolare started in the autumn of 1992. The contact had been established through Crystalox, which at the time had a close cooperation with Eurosolare for the development and commercialization of Bridgman crystallization machines. Crystalox was owned by Elkem since 1985 as a consequence and heritage of the Exxon project (1980–1985). From its position as an equipment supplier of specialized devices to academic research, Crystalox wanted to grow its business as a supplier to the PV industry. Multicrystalline silicon was recently recognized as a good compromise between the expensive single (or monocrystalline) silicon and the cheaper but unstable and far less efficient amorphous silicon. Several companies, for example, Kyocera in Japan, Solarex in the United States, and Photowatt in France bet on this technology and invested aggressively in it. These companies had to design and construct their own crystallization machines. Crystalox foresaw therefore a major business opportunity in commercializing Bridgman machines for the promising PV market. The company was founded in 1982 by the Oxford professor Dr. David Hukin. Selected as a subcontractor to the Elkem–Exxon project, it had been acquired by Elkem in 1985 just before the early termination of the project. After 1985, Crystalox remained the sole PV activity within Elkem until approximately 1990. At the end of this 5-year long walk through the desert, Crystalox brought to its Norwegian mother company valuable contacts among customers, who wished Elkem on board in developing a silicon grade for solar cells, which could not be found in the market (see Box 4.1). They had several good reasons to approach Elkem. The company was highly respected as a global leading producer of metallurgical silicon (MG-Si), Elkem was

BOX 4.1   CRYSTALOX: DATES AND EVENTS

Crystalox was founded in 1982 with headquarters near Oxford in the United Kingdom. The company was able to draw on a combination of engineering and materials science expertise to develop R&D systems for purification and crystal growth of metals, oxides, and semiconductors. Elkem first collaborated with Crystalox in 1984 on a joint project with Exxon Solar Power Corporation to purify silicon for PV applications.

This cooperation eventually led to Elkem's takeover of Crystalox in July 1985. While work continued on the silicon purification program, the company also continued its production of crystal growth systems and supplied equipment to customers in universities and research institutes in Europe, the United States, and Asia.

In 1990, the company started its pioneering development of systems for industrial production of multicrystalline silicon ingots for the nascent PV market. The induction-heated systems used the directional solidification process to crystallize silicon in 440 mm square silica crucibles with an initial ingot size of 70 kg suitable for producing a 4 × 4 matrix of 100 mm square blocks compatible with the industry standard 100 mm wafer at that time. Wafers produced using the directional solidification process offered superior performance to those produced using the casting process, which had been employed in the industry until that time due to the better impurity segregation and columnar grain growth.

The first commercial systems were supplied to a company in Italy and the crucible size was soon after increased to 550 mm square. Thereafter, the process gained widespread acceptance within the growing PV industry and subsequent production systems were supplied to customers in France, Germany, and Japan. The directional solidification process using silica crucibles remains the PV industry standard for the production of multicrystalline silicon to this day.

However, despite Crystalox's success, Elkem wanted to focus on its core business interests in ferrosilicon, carbon, and microsilica and sold the company to the management in an MBO in December 1994.

Under its new ownership, the company continued its technology development and made rapid progress in scaling up the process. In 1996, the next-generation system capable of producing 66 cm square ingots was installed at a customer's site in Japan. These 250 kg ingots were sized so that they could be used producing a 5 × 5 matrix of 125 mm blocks or a 4 × 4 matrix of 156 mm blocks. The 156 mm multicrystalline silicon wafer has since then become the workhorse of the PV industry.

As part of the development program, Crystalox was increasingly producing multicrystalline ingots in order to optimize the system performance. Following the encouragement of its world-leading customers, the company set up its own commercial ingot production facility in 1998. Ingots were primarily shipped to Japan, which at the time was the major manufacturing center for the PV industry. Ingots were converted to wafers in partnership with several subcontract wafering companies, which were beginning to operate in Japan.

Germany was also becoming an increasingly important PV manufacturing center and PV Silicon was founded in 1997 in Erfurt, Germany, initially to produce monocrystalline silicon wafers. In cooperation with an equipment manufacturer, the company developed the first generation of wire saws. In order to serve the growing demand for wafers in Germany, Crystalox entered into a strategic partnership with PV Silicon in 1999 to produce multicrystalline silicon wafers.

As the relationship between Crystalox and PV Silicon strengthened, the two companies merged to form the PV Crystalox Solar Group in 2002. In the same year, the sales division Crystalox Japan K.K. (name subsequently changed to PV Crystalox Solar K.K.) was founded to further improve relationships with customers in Asia.

In June 2007, PV Crystalox Solar made a successful flotation on the London Stock Exchange, raising additional funds to further expand its operations and international business. The company continued to concentrate ingot production in the United Kingdom and to convert the ingots into wafers at its own facilities in Germany and its subcontract partners in Japan. By 2012, the company had a production capacity of 750 MW.

The collapse in PV market pricing, which started in 2011 due to industry overcapacity primarily in China and which persists to this day, led many companies to exit the PV industry either voluntarily or through insolvency. In Europe, most commercial wafer and cell production was shut down but Crystalox maintains its production operations in the United Kingdom and Germany albeit with output at a lower level. Wafer shipments of around 210 MW were achieved in 2014 primarily to customers in Taiwan.

Iain Dorrity
Private communication, November 2015

known to possess a solid technology platform, and to have a significant industrial capability with plants in Norway and North America. The Exxon–Elkem project was fresh in everyone's memory. Its results had been broadly disseminated by the U.S. DoE, which had triggered and partly financed it through the JPL-LCSSA project. Elkem also had close contacts and discussions with a similar Japanese New Energy and Industrial Technology Development Organization (NEDO) program, headed by Professor Sano, a steel and ferroalloys metallurgist. Silgrain technology and the marketing of its product as raw material to the electronic-grade silicon—polysilicon—also contributed to the common belief that Elkem had all keys in hand to develop a solar-grade route by means of metallurgical processes. Therefore, companies like Eurosolare approached Elkem through Crystalox and through Elkem's international sales network, particularly the regional offices in Tokyo and Düsseldorf.

It was in this context that Eurosolare in the course of 1992 invited Elkem to a joint project with the objective to develop an SoG-Si to be used as feedstock for multicrystalline ingots in Bridgman furnaces originally designed and developed by Heliosil (see Chapter 2) and further manufactured by Crystalox. Eurosolare (formerly Italsolar) was a pioneer of the PV European industry. It was wholly owned by ENI, the Italian state energy and utility company, which also owned the petroleum company AGIP, including its subsidiary Norsk AGIP operating on the Norwegian offshore oil fields in the North Sea. According to the governing rules in force at that time, to be granted exploration and exploitation concessions by the Norwegian Department of Energy (OED), oil companies had to contribute to land-based not petroleum-related R&D activities benefitting national industry and employment (see Section 4.4.1 about Exxon). During the autumn of 1992, Elkem, Crystalox, and Eurosolare entered into a 3-year cooperation consortium agreement with Norsk AGIP; the latter would contribute the major part of the financing. Also, the research council of Norway would make a significant financial contribution. Elkem would fine-tune its metallurgical process to solar quality at affordable cost, Crystalox would further develop its Bridgman furnace (increasing the size was essential to achieve higher productivity in ingot manufacturing), and Eurosolare would make wafers and cells and characterize all products in terms of chemical analysis, and electrical and other physical parameters. Reference ingots made of electronic-grade polysilicon as well as synthetic feedstock simulating the purity achievable with the Elkem process were also included in the program. The target for Elkem was to achieve cost below $20/kg for the purity specifications listed in Table 4.3 [48].

These characteristics were defined as Elkem's best capability, deduced from the former project with Exxon and from quantitative modeling by the company's experts. The intended product was given the trade name SolarSil™ and was subsequently described and published in a review paper [49]. With funding from Norsk AGIP, a 100 kW induction furnace for silicon refining was purchased (from Inductotherm) and installed at Elkem Research in Kristiansand. It was proposed to generate 100 kg of SolarSil as an important milestone to demonstrate the process.

Under this program, the technical team at Elkem Research carried out numerous basic experiments that would prove to be of great value 10 years later under the Elkem Solar building era. The contemplated process built on the main principles

TABLE 4.3

Tentative Product Specification for Elkem–Eurosolare Project

Impurity	Tentative Product Specification (ppmw)	Detection Limit		Accuracy	
		Strongly Preferable (ppmw)	Minimum Required (ppmw)	Strongly Preferable (ppmw)	Minimum Required (ppmw)
B	<0.4	0.1	1	±0.1	±1
P	<0.7	0.1	1	±0.1	±1
Ca	<5	0.1	5	±0.1	±3
Other metals, each	<1	0.1	2	±0.1	±2
Other metals, sum	<5				
C (total)	<10	1	10	±1	±10
O (total)	<50	1	50	±1	±30

Source: Adapted from A. Schei, 1993. Equipment for Eurosolare program, *Memo, 3.05.1993.*

as in the Exxon study. However, the concept of using very pure raw materials in the carbothermic reduction was abandoned. The team put stronger emphasis on the post-furnace refining by both pyro- and hydrometallurgical techniques and carried out systematic studies on

- Slag refining (solid–liquid and liquid–liquid extraction of impurities)
- Gas bubbling (liquid–gas extraction of impurities)
- Filtration of liquid silicon
- Decantation of solid impurities in liquid silicon
- Alloying of silicon to dissolve impurities in leachable phases
- Leaching
- Casting and solidification

These techniques and the potential they revealed under the experiments were benchmarked against results, publications, and ideas brought to Elkem by an increasing number of researchers, inventors, and newly inspired entrepreneurs. During this program and under these circumstances, the Elkem technical team started to look at a continuous refining process, which was already known from an Australian project for refining aluminum metal [50–53].

The Eurosolare project was under the formal responsibility of the silicon metal division of Elkem but was driven by the corporate technology function. The program had been initiated and designed by Elkem's CTO (Dr. Alf Bjørseth), who through his position in the Crystalox' board of directors, had better insight into this business than anyone else in the corporate management. The technical results achieved under this program were also very important for the Texas Instruments project. This latter project was run in parallel with the Eurosolare and used the same resources, but had higher priority within the silicon metal division, which had launched it.

4.5.2   Texas Instruments: Silicon to Spheral Solar Cells

Texas Instruments (TI), the North American semiconductor producer, who worked on a new solar cell concept, approached Elkem's silicon division through other channels than Crystalox. In TI's concept, the cells consisted of tiny crystalline silicon spheres (1–2 mm in diameter) embedded into flexible aluminum foil [54]. TI aimed at low-cost solar panels, which could easily be applied in a multitude of applications demanding a flexible substrate. This technology became better known some 10 years later when it was acquired by ATS, an Ontario investor, and further inspired a couple of similar concepts ("spheral" or "spherical cells") by companies in Japan. ATS, who at the time also owned Photowatt in France, built a 20 MW plant in Waterloo (Ontario) around 2002–2006 to introduce the spheral cells to the market. However, the plant was never fully commissioned. Our understanding is that several steps of the mechanical character failed during the upscaling and could not be solved without significant redesign of the process and the plant. Changes at ATS independent of this project may also have contributed to loss of confidence and interest in spheral cells technology. ATS stopped the project and later withdrew from Photowatt as well. But in 1991, the developers at TI strongly believed in the concept, which they considered a potential breakthrough technology. A benefit of the new technology was to refine the silicon when producing the cells. They believed the silicon feedstock could be of metallurgical grade, which would cost 5%–10% of polysilicon or downgraded electronic silicon, then the current feedstock to solar cells. However, TI specifications required smaller donor and acceptor concentrations than most grades of metallurgical silicon have. They required B < 5 ppmw and P < 10 ppmw, but in most metallurgical grades of silicon, these elements have a concentration above 20 ppmw. The TI process could tolerate the metallic impurities at the same level as in the commercial grades, but the lowest possible concentration (especially of Fe) would clearly be an advantage. Carbon (C), which is present at a level of several hundred ppmw, soon became a technical issue for TI. The silicon feedstock should also be shaped as tiny particles in a narrow distribution between 1 and 2 mm to achieve a homogeneous distribution of cells. The Elkem product dedicated to TI was designated as SolarSil B or SolarSil AT2. The Elkem team had in mind the promising results from the joint project with Exxon and knew that with the existing equipment and plants, Elkem should be capable of satisfying TI's demand. In usual North American fashion, aggressive market scenarios were put forward requiring thousands of tons of SolarSil AT2 at a time when other customers such as Bayer Solar or Eurosolare talked in best case of tens of tons. This seemed an attractive business case for Elkem's silicon division and especially its North American plant at Alloy (West Virginia), which by all means fought to gain the business with TI. The project was soon coordinated by a North American project manager whereas the technical development and most of the trial production of SolarSil AT2 were carried out at Elkem Research in Kristiansand and Elkem Bremanger (the Silgrain plant) using the same professional resources as the Norsk AGIP/Eurosolare project for SolarSil AT1.

The trial production was to a large extent financed by TI, but benefitted also from the more generic PV projects on SolarSil AT1, which had been launched by the silicon or technology divisions of Elkem with partial financing from Norsk AGIP and the Norwegian Research Council (NTNF). To demonstrate the process on a large scale, Elkem purchased some hundred kilograms of a low-boron (B < 5 ppmw) MG-Si produced by Kemanor, a Swedish MG-Si producer. This material was left-over from R&D trials conducted in the 1980s in cooperation with Heliotronic (a subsidiary of Wacker, in Germany, dedicated to SoG-Si). In 1991–1992, a 20 kg batch of SolarSil AT2 was produced at Elkem Research and Elkem Bremanger applying slag treatment in an induction furnace and acid leaching similar to the Silgrain technology. Based on this campaign, the silicon division carried out a cost estimate on capital and operational expenditures for both SolarSil AT1 and AT2 for 1000 MT/year of these grades. In 1993, a larger batch of 30 MT crude silicon from Elkem Alloy (West Virginia) was slag treated at Elkem Research and prepared for leaching at the Silgrain plant (Elkem Bremanger). To our best knowledge, the entire batch had not been completely leached when the project suddenly met its end in 1993. Nevertheless, sufficient data had been collected to demonstrate the capability of the process.

As mentioned earlier, the particle size distribution was an important specification expressed by TI as all spheral cells should be the same size. Silgrain technology, which achieved higher purity with respect to phosphorus, iron, and other transition elements, was also an interesting technology for TI as it generated small particles with a relatively narrow size distribution, which by means of screening could meet the TI specification. During the autumn of 1993, TI and Elkem discussed the licensing of Silgrain technology, initially for research purposes. Other fractioning and sizing techniques, for example, atomizing and granulation were proposed and discussed without a deep involvement from Elkem as TI claimed to have its own methods.

At the end of 1993 and early 1994, TI decided suddenly to stop the project. The TI team was dismantled and the spheral cell technology was sold to a small group of entrepreneurs with background from the project. Later, these entrepreneurs sold the project to ATS, an Ontario-based technology investor, who made a fair attempt at industrializing the process (see earlier), however, without Elkem's contribution as feedstock provider.

4.5.3 Development Slows Down Again (1994–1996)

The two projects (Norsk AGIP/Eurosolare and Texas Instruments), which were launched in 1991–1992 and lasted until 1994, represented a "renaissance" of interest in PV within Elkem after the Exxon project had faded out. Ahead of these two projects, the silicon and technology divisions had revisited their 5–10-year-old files and carefully reconsidered a modest program on PV encouraged by hints from Crystalox and other contacts in the market as well as stronger incentives from the Norwegian Research Council, which could finance up to 50% of the R&D efforts. During this time frame, the 100 kg SolarSil AT1 target had been decided and in 1991 Elkem hired a marketing director to work out a solar strategy, and coordinate and lead all photovoltaic activities within the Elkem group. Increasing interest and growing activities were indeed spread out among a plethora of units, that is, the technology and the silicon divisions in Oslo, the Bremanger plant on the Norwegian west coast, the research center in Kristiansand on the Norwegian south coast, the stand-alone company Crystalox in the United Kingdom, the sales offices in Japan and Germany, and an unofficial North American division in Pittsburgh, which, thanks to a historical plant (Alloy in West Virginia), a customer base, and rich history from its time as part of Union Carbide, wanted to play a first-in-the-class role in this new business. In November 1991, the management of the silicon division adopted an ambitious strategy. The vision was to become the leading global supplier of SoG-Si, ingots, and wafers with a market share of at least 80% by 2000. This position was to be built on a technology platform consisting of Elkem's unique feedstock technology and Crystalox's further developed Bridgman technology.

However, this renaissance came at an unfortunate time for Elkem. The company encountered hard times with heavy operational losses as a consequence of a negative international economic cycle and major geopolitical changes (the first Gulf War, the Soviet Union's collapse, downturn in the offer–demand cyclic phases). The company would have entered into bankruptcy if the Norwegian government had not rescued it and guaranteed its survival. Elkem had at that time all technical and industrial assets to quickly succeed in the PV business and take a leading role, but the company lacked the necessary financial strength.

The projects had to be financed with great creativity mainly through contributions from partners and customers (Norsk AGIP and Texas Instruments) and public support (NTNF). The guidance from the management was that everything needed to be externally financed. At the end, the management lost confidence in the company's ability to realize its ambitious vision. In the middle of 1993, the ambition was considerably reduced as expressed in the directives from the silicon division. After the termination of the project with TI, the technology division was also instructed to terminate the Eurosolare project and the contract with Norsk AGIP as soon as possible and no later than by the end of the first quarter of 1994. The Norsk AGIP contract was entered into for a period of 3 years and was already capitalized for the first 2 years (1993–1994), but was suddenly adjourned unilaterally by Elkem and never conducted to its end.

In 1994, following the same strategic pattern, the silicon division was recommended for withdrawal from Crystalox, leaving it to itself through a *management buyout* (MBO) transaction. The rationale behind the decision to withdraw from these projects was more the consequence of a business reengineering process within Elkem, realigning on its fundamentals, than a cost-saving measure. Although these projects were all externally financed, they bound valuable resources and competences that the group wanted to reallocate to activities closer to the assumed core businesses. Frequent changes in the management at all levels of the company (corporate, business units/divisions, operational units/plants) did not create the appropriate conditions for a long view strategy.

However, it did not last long before new signals came from the executive management itself. The CEO (Ole Enger) showed deep respect for the unique technology competence in the company. Moreover, the outstanding results achieved during the solar research had a major influence on the understanding of the silicon process and products. The silicon business, in spite of poor financial performance, became in this period a business area with increased attention from the corporate management. A good example of this is the Refining Competence Center, which the silicon division founded in 1994 and located as a department of the research center in Kristiansand. The goal was to gather at one place significant resources dedicated to collect systematically all knowledge and experience related to pyrometallurgical refining of silicon regardless of the application. Most of the experts associated with this center had invaluable experience and background from the solar projects. They could take care of a smooth technology transfer to other fields of commercial application, that is, silicon to the chemical industry (silicones and polysilicon), which had the highest priority and silicon to aluminum alloys (pulled by the demand from automotive, construction, and packaging industries). Also, the application of ferrosilicon in steel and iron alloys foundry benefitted from the systematic knowledge and competence built at this center.

In the 1994–1996 period, the modest activity related to solar was incorporated in the Refining Competence Center with strong focus on boron removal and composition of silicon alloys suitable for leaching. The activity was mainly technology driven and benefitted from funding from the Norges Teknisk Naturvitenskapelige Forskningsraad (NTNF).

4.6   THE RENAISSANCE (1996–1999)

4.6.1   Reentering the PV Business through European Research Consortiums

Early in 1996, Elkem started the work to set up a European research consortium, which could partly be financed by the European Commission (EC) within the JOULE program framework. This preliminary work was carried out by the head of Elkem Research, Dr. Knut Henriksen, with the technical support of Dr. Cyrus Zahedi. The idea was to make use of upgraded metallurgical-grade silicon (UMG-Si) similar to SolarSil, or even a lower grade of SolarSil, to produce flat substrates on which a thin film of multicrystalline purer silicon would be grown by *liquid-phase epitaxy* (LPE). In this technique, the thick solid substrate of (upgraded) metallurgical silicon is exposed for a short period of time to a liquid bath of molten pure silicon (e.g., electronic grade). A thickness of a few tens of micrometers for the thin layer is sufficient to assure the p–n junction required for the photovoltaic effect. The thick substrate with its thin layer of pure silicon is considered a *wafer equivalent* in principle able to substitute a multicrystalline silicon wafer. Such a technique to produce wafers would considerably reduce the amount of hyperpure silicon (electronic grade) needed to produce solar cells and thus create a market for UMG-Si, which was within Elkem's industrial capability. The other partners in the consortium were Eurosolare (Italy), Fraunhofer Institute ISE (Germany), University of Konstanz (Germany), and the IFE, Norway. Elkem had the honor to head and coordinate the consortium. The project known under the acronym SCARF started during the autumn of 1996 and lasted for 3 full years. Trial production of UMG-Si was performed at Elkem Research and the LPE deposition at the Norwegian institute IFE, with whom Elkem entered into a bilateral cooperation agreement. Until then, IFE had played a minor role within PV. Lifted by this project, IFE eventually built a significant competency and capability within the photovoltaic area and was later assigned the role of coordinating all Norwegian publicly funded PV research.

During the same period, Elkem, through its silicon division, was invited to a competing EC JOULE project under the acronym MAGSIFIC, which differed from SCARF in the way the thin pure crystalline layer was grown. CVD using trichlorosilane as a silicon precursor was a well-established method in semiconductor manufacturing. MAGSIFIC proposed to make use of this readily available technique to depose the thin layer on the UMG-Si substrate. The other steps and goals pursued by MAGSIFIC were quite similar to those of SCARF. The other participants in MAGSIFIC were BP Solar (coordinator, UK), Crystalox (UK), Fraunhofer Institute ISE (Germany), and the Belgian institute IMEC (Belgium).

Both projects generated significant knowledge for Elkem and the other participants. Beyond technology achievements, they contributed to develop a broader market knowledge and to build strong relationships at the European level, which should prove to be of significant value for the next decade. The technology concepts explored by these two consortiums were never implemented as such by any of the partners. However, at the time of writing, several startup companies, particularly on the U.S. West Coast, are pursuing similar concepts. Recently the Fraunhofer Institute ISE, which participated in both consortiums, has spun off its long-lasting activity on *wafer equivalent* into a dedicated startup company NexWafe GmbH.

SCARF and MAGSIFIC were the two first projects in a long list of EC-supported PV projects with Elkem participation.

4.6.2   THE GREAT RENAISSANCE

The great renaissance of PV at Elkem was launched at the initiative of the executive management itself around 1997. The company had gone through a successful turnaround operation partly triggered from the inside, but partly engineered by external consultants (Boston Consulting Group). Sales prices for the company's main products (aluminum metal, ferroalloys, and silicon metal) had recovered to a viable level, thus enabling a decent financial recovery. The executive management (CEO Ole Enger) was shown unanimous great respect and confidence by employees and shareholders. The management could therefore start thinking about growth and planning strategic moves. With an annual growth rate of more than 4%, silicon metal was the business unit at Elkem, which offered the most potential growth opportunities. The corporate management asked the silicon division to perform a study to identify a set of promising business opportunities and to propose a road map for a downstream strategy building on the company's position as a silicon metal manufacturer.

In 1997, the solar cell market was still very modest (115 MW), equivalent to around 2000 MT of pure silicon. However, the two-digit growth rate for several years in a row was well above 20%. The growth was stimulated by incentive programs, first of all the *rooftop program* in Japan. Beyond this temporary incentive from a national program, one could feel a real aspiration for what today is called the *energy transition*. Soon after, the Japanese incentive program was overtaken by a singular strength by the *feed-in-tariff* (FIT) program in Germany. Long-term forecasts and scenarios by energy companies and market analysts all pointed out a bright future for solar energy. Several materials were suggested to convert light into electricity. But silicon, which had acquired a solid reputation as semiconductor material, had already established hegemony with a market share above 90%, including amorphous silicon. No material seemed capable to threaten silicon in the predictable future. Were the forecasts right, the silicon market for solar cells would grow from a few hundred metric tons per annum to several hundred thousand tons within a 10–20 year time frame. Polysilicon production via the decomposition of chlorosilane in a bell-jar reactor (Siemens process) seemed at the time too complex, too energy consuming, too difficult to scale up, and hence too expensive for the mass development of solar energy. All prognoses pointed out challenges and opportunities, which the executive management at Elkem wanted to understand better.

Meanwhile a former Elkem's CTO Dr. Alf Bjørseth had left the company and founded ScanWafer AS (1994), a new venture aimed at producing and marketing multicrystalline silicon ingots and wafers for the solar market. The company was located at an industrial park of Norsk Hydro (later Yara) in a remote place in a fjord just north of the Arctic Circle (66 N). Settling there brought a significant incentive package from the local industrial site owner as well as from the municipal, regional, and national authorities. Starting from scratch, the new venture planned to use commercially available raw materials (polysilicon of various grades suitable for solar cells) and equipment, that is, Bridgman furnaces. The production started in 1997 with two furnaces supplied by Crystal Systems (USA). Expertise and experience from the past Elkem–Crystalox–Eurosolare cooperation must have been an essential asset for the new venture. Besides the founder Dr. Alf Bjørseth (former Elkem CTO and chairman of the board of Crystalox), two prominent experts from the past cooperation took an active part in working out the technology platform for the company, selecting the equipment and the raw materials, following up the construction, writing the operational procedure, training the operators and the technical management, and following up the quality improvement several years after the company was founded. These two experts were Dr. David Hukin, founder and former general manager/CEO of Crystalox, and Dr. Daniele Margadonna, former CTO at Eurosolare.

In 1997, 3 years after exiting Crystalox, Elkem invested in ScanWafer, acquiring slightly less than 10% of the share capital. With this transaction, Elkem showed to the world its renewed ambition to take part in the emerging, fast-growing solar cell industry. As part of the shareholder agreement, the two companies (Elkem and ScanWafer) launched a joint technology program aiming at developing silicon processes suitable and dedicated to the production of solar cells. Purity requirements that ScanWafer imposed on silicon feedstock were, however, closer to the electronic-grade than the best-grade Elkem was capable of. ScanWafer wished a simplified polysilicon process whereas Elkem wished a process based on metallurgical refining steps. Therefore, the cooperation soon became less fruitful and harmonious than first anticipated. Both companies after a while pursued their own separate ways to SoG-Si and it would take another 5–6-year period (2004) until they would find a common ground when Elkem and its owner, the Orkla group of Norway, massively invested in REC, the integrated photovoltaic industrial group that had evolved from ScanWafer (see Box 4.2 about ScanWafer and the REC group).

With the great renaissance for solar cells, Elkem gradually increased its R&D activity and budget in this field. A large part of the activity was conducted in *terra cognita* on slag refining and leaching. Identification, search, and cleaning of

BOX 4.2   REC: DATES AND EVENTS (1994–2015)

Ingot and Wafers

1994: ScanWafer AS was founded at Glomfjord (Norway) by Dr. Alf Bjørseth and Reidar Langmo.

1997: ScanWafer started the production of multicrystalline wafers, two furnaces. Capacity: 2 MW.

1997: ScanWafer decided to expand capacity with eight additional furnaces, capacity: 10 MW. Elkem entered into ScanWafer shared capital at close to 10%.

1999: SiNor AS started the production of monocrystalline silicon ingots for semiconductor market.

2001: New ScanWafer plant (SW II) at Glomfjord. Capacity: 40 MW initially; then second phase to 350 MW total.

2003: New ScanWafer plant at Herøya (SWH I). Capacity: 200 MW.

2007: New ScanWafer plant at Herøya (SWH II). Capacity: 200 MW.

2008: New REC Mono plant at Glomfjord. Capacity: 250 MW.

2010: New ScanWafer plant at Herøya (SWH III–IV). Capacity: 400 MW.

2006–2010: Planning, construction, and implementation of REC plant in Singapore for the production of ingots, wafers, cells, and modules. Integrated capacity (modules): 800 MW.

Cells and Modules

2000: ScanCell AS in Narvik (Norway) and ScanModule AB in Arvika (Sweden) were established as business entities for manufacturing cells and modules, respectively. The idea was to use wafers from ScanWafer. Renewable Energy Corporation AS was established as a holding group, including assets in ScanWafer, ScanCell, and ScanModule.

2005–2010: Successive expansions at ScanCell 200 MW (Narvik) and ScanModule 200 MW (Arvika).

2006–2010: Planning, construction, and implementation of REC plant in Singapore for the production of ingots, wafers, cells, and modules. Integrated capacity (modules): 800 MW.

Silicon Feedstock

2002: REC and ASiMI/Komatsu formed Solar Grade Silicon LLC (SGS), a 50–50 joint venture company at Moses Lake (USA) for polysilicon supply to ScanWafer and development of the innovative FBR polysilicon process.

2005: REC acquired the whole ASiMI/Komatsu polysilicon business including all shares in SGS, two production plants in the United States (Moses Lake and Butte), and all associated technology and business (silane, Siemens with silane, and FBR with silane). Capacity: 5000 MT polysilicon with silane Siemens.

2005–2009: Planning, construction, and implementation at Moses Lake (USA) of silane-FBR with brand new FBR technology. Capacity: 9000 MT granular FBR polysilicon.

Company Structure

1994: ScanWafer AS was founded at Glomfjord (Norway) by Dr. Alf Bjørseth and Reidar Langmo.

1997: Elkem entered into ScanWafer share capital at close to 10%.

2000: (Re)-foundation of Renewable Energy Corporation.

2005: Elkem acquired approximately 20% of REC.

2006: Successful IPO of REC ASA at Oslo stock exchange.

2008: The global financial crisis has a strong negative impact on financing the emerging and fast-growing PV business. Linked together by supply–purchase and/or technical cooperation agreement, all PV companies were dependent on each other. The fall of one affected the other. REC was strongly affected.

2010: Orkla (Elkem's sole owner and publicly listed company at Oslo stock exchange) increased ownership in REC controlling with Elkem 40%.

2011–2012: All REC plants in Norway (ScanWafer and ScanCell) and Sweden (ScanModule) were closed down and the REC subsidiaries in these countries put in for bankruptcy. REC ASA still listed at Oslo stock exchange continued with assets in Singapore (the new plant for wafer to module) and in the United States (two polysilicon plants and the brand new FBR technology).

2013: Fission of REC ASA into REC Silicon ASA and REC Solar ASA, both listed on the Oslo stock exchange. REC Silicon ASA owned the silicon plants in the USA and associated business; REC Solar owned the wafer to module plant in Singapore and all associated business.

2014–2015: Bluestar-ChemChina acquired REC Solar ASA with the intention to merge with Elkem Solar. Elkem Solar announced the takeover of the REC ingot plant at Herøya from the bankruptcy estate, planning refurbishing and startup in 2016.

slag-forming additives were given high attention and priority as availability of such materials was perceived as a strategic issue. New silicon alloys suitable for leaching were systematically investigated. These studies resulted in several patents and paved the road to the future process, which would be industrialized a decade later by Elkem Solar. However, the gap was still large between Elkem's best capability and the requirements/expectations from the increasing number of potential customers visiting Elkem Research in Kristiansand (where most of the research took place). The concept of *compensation* between *donors* and *acceptors* was not yet fully known or accepted in the industry. Having doubts, executive management hesitated to follow the experts' recommendation to put all efforts on the route, which had been formerly explored through the Exxon and Eurosolare projects. The technical team assessed therefore alternative routes, which came up through the literature or were introduced by inventors or business partners. Examples were plasma refining techniques and continuous crystallization techniques, which both were the subject of numerous research projects worldwide (see Chapter 3). In this period, the technical team together with the Norwegian research contract foundation SINTEF investigated a continuous crystallization technique, which was known through a research project by CSIRO in Australia for refining aluminum metal [50–53]. Its application to silicon seemed more complex because of the higher melting point (1415°C for silicon versus 660°C for aluminum). Taking necessary adjustments into account, the Elkem–SINTEF team developed a model, which showed promising refining potential, but also high risks on the equipment design and material selection. The process was internally designated SIMIG (for silicon refining by migration of impurities). Assessing its application to silicon purification unfortunately did not go further than modeling at the pure conceptual level.

Under the great renaissance, Elkem cultivated contacts with many groups and companies in the United States (AstroPower), Japan (Kyocera, Sharp, Mitsubishi/Melco), Europe (BP Solar, Bayer Solar, Fraunhofer Institute ISE, University of Konstanz), and Norway (ScanWafer, IFE, SINTEF, NTNU) to name a few among them. These relationships would strongly influence the company's attempt to play a significant role in the solar industry.

4.7   ELKEM SOLAR, FROM PROJECT TO BUSINESS UNIT (2000–2006)

4.7.1   Elkem Solar: An Elkem-Driven Initiative

Since the 1980s, Elkem has gradually decreased its external engineering services. From being one of the leading suppliers of furnace technology, Elkem has become a global ferroalloy producer and the engineering department at Elkem has been reduced from approximately 250 persons in the 1980s to an internal support staff, which today (2015) counts 20 people. This change also influenced the technical activities and capabilities at the corporate R&D center located in Kristiansand. For 50 years, this well-equipped center had been used to test new equipment, raw materials, and process alternatives for external technology sale. A large R&D organization had been established and pilot facilities, including workshops and laboratories, supported the experimental work. Under the new industrial regime, the R&D capability was redirected toward internal support to the company's plants and business divisions keeping the employment at the center at a high level regardless of the top and down fluctuations affecting the overall company's performances. The capability due to unique competence and equipment at the corporate R&D center will turn out to be an invaluable asset for further solar development at Elkem.

At an internal strategy meeting in 1997 revisiting SoG-Si opportunities [45], recent contacts with possible partners were discussed. Among the companies in question were Bayer Solar, Sumitomo, AstroPower, Sharp, Texas Instruments, and also the newly established Norwegian company ScanWafer. These companies represented different technologies and the seminar concluded that Elkem's options to enter the field seemed confined to three possible approaches:

- Upgraded metallurgical silicon (UMG-Si)
- The migrating crystal process SIMIG
- Simplified production of polysilicon

In a later study, the consultant company McKinsey & Company [55] recommended strategic moves for improved utilization of the technology environment in Elkem. The following advice influenced the establishment of Elkem Solar:

- Elkem should strive to be a leader on product development in silicon.
- Elkem should manage (and fund) growth/business building projects separate from process/product development programs.

With the wide experience from earlier solar-related projects and advice from both internal strategy work and external consultants, Elkem Solar was established as a project in 2001. The project was lifted from the silicon division as an entity reporting to a steering committee representing the corporate and technology management.

4.7.2 COOPERATION ELKEM/ASTROPOWER: A MUTUAL ADVANTAGE (1998–2003)

The migrating crystal process (called SIMIG in Elkem; see Section 4.6.2 and References 50–53) had been investigated in Elkem in the late 1990s as an alternative refining process for silicon in parallel with new evaluations on both slag refining and leaching. Elkem had been part of two European projects on wafer production by epitaxial growth on cheap substrates—SCARF and MAGSIFIC. UMG-Si was used as substrate. At an Elkem internal workshop in 1999 [56], reference was made to a European white paper pointing out a future trend toward continuing dominance of the crystalline silicon wafer technology and the need to develop low-cost SoG-Si. The existing producers of electronic-grade silicon showed no interest in downgrading their high-value product for the solar market, but Elkem saw a possibility thanks to its background in the early solar projects. Elkem needed yet a downstream cooperation partner to validate its strategy.

For Elkem's office in Tokyo, sourcing of waste materials from monocrystalline silicon production in Japan was part of the business for several years. The sourcing activity also included scrap wafers from the electronic industry. AstroPower in the United States, then a leading PV listed company, had developed a process where the printed devices on the recycled wafers were removed and the cleaned wafers were then used in solar cell production. As Elkem-Japan supplied wafers to AstroPower, a close relationship was built up between the two companies, and this also included research people involved with UMG-Si at Elkem's research center in Kristiansand. The common interest grew into a R&D program where Elkem would prepare and supply silicon from internal SoG-Si campaigns and AstroPower would use such material in its proprietary Continuous Uni-Directional Solidification (CUDS) technology and Silicon Film™ technology. The cooperative partnership was formally established on December 6, 2000 by signing a technology cooperation agreement (TCA) [57].

Elkem had for many years supplied silicon to the polysilicon producers (Siemens process) from the Silgrain plant in Bremanger, Norway, and was therefore familiar with the purity requirements of both the electronic and solar industry. A production technology based on pyro-metallurgical refining was not believed to satisfy the required impurity limits. In or around 2000, AstroPower was a highly respected company and was therefore a perfect partner for Elkem. At a meeting in Delaware in October 2002 [58] where the executive managers of both companies were present, the purity specification and the demand for raw material to the CUDS process were ceremoniously agreed upon. For the doping elements (B and P), 5 and 10 ppmw (respectively) were defined as upper limits and a total of 500 ppmw for all other impurity elements was accepted (far from the 0.1 ppmw total impurities in polysilicon from the conventional Siemens process). Although the CUDS technology was AstroPower's intellectual property, AstroPower wanted Elkem to incorporate it into its SoG-Si process as a solidification process. In return, Elkem should secure AstroPower's long-term plans with an annual supply of at least 6000 MT silicon feedstock.

The wafers from the Silicon Film casting machine were at the beginning 600–800 μm thick, meaning twice the thickness of wafers from standard commercial wafer lines. This was gradually improved, but the technology did not allow wafers thinner than 300–500 μm. In principle, this would also limit the use of feedstock with low lifetime. However, the cost per energy unit installed was still favorable because the CUDS technology had both low *opex* and *capex*. During 2002, a "close to industrial scale pilot unit"—also called Recycle Machine—was under construction and mechanical work started in May the same year [59]. However, the progress was slow and in January 2003, the Recycle Machine was only 80% complete [60]. In June 2003, the unit had still a way to go before completion. From the quarterly reports published in the same period, it became obvious that the listed company AstroPower faced serious financial problems [61], and in October the same year (2003), Elkem decided to terminate the TCA of December 6, 2000 [57].

Results from the Silicon Film trials were not so promising compared to other results Elkem had obtained and would publish in January 2004 [62]. The silicon was MG-Si purified by slag treatment, leaching, and solidification produced in a small pilot at Elkem's research center. The produced SoG-Si was then solidified at SINTEF in a Bridgman crystallizer (purchased from Crystalox), and cells were made at the University of Konstanz (Germany). Elkem also developed its own DS technique with the FFS (furnace for solidification) (see Section 4.8.5). Results from Silicon Film cells had given 9% efficiency with both pure polysilicon and Elkem's SoG-Si, while a level of 14%–15% [63] was obtained when the FFS crystallizer was used.

During the following years of Elkem Solar's development, the FFS technology was further improved and the only applied technique for DS in Elkem's process chain to SoG-Si. However, other crystallization/solidification techniques were tested. For instance, the Ukrainian company Pillar had developed a modified continuous solidification unit and invited Elkem Solar to take part in the development of the concept [64]. Another concept based on "Top Grown multicrystalline Solidification" (TGS), was under development together with the FCT Systeme GmbH. Both these projects aimed at simplifying the process value chain (see Section 4.9.4), but were later abandoned as other technologies proved more cost efficient.

4.7.3   Two Supply Contracts in Place Ahead of Industrialization

4.7.3.1   Cooperation with BP

In January 2003, BP Solar and Elkem entered into a confidentiality agreement and started common work from the autumn of 2003 when BP Solar started to test feedstock from Elkem's pilot production. The initial results were promising and in June 2004, the two parties signed a cooperation agreement. The main objective was to verify the SoG-Si quality in solar cell production at BP Solar [65]. BP Solar used standard solidification technology for their multicrystalline ingot production and therefore expressed standard quality requirements on feedstock far more demanding than AstroPower's. These major differences led to the test production of two different products: one for AstroPower (until termination of the TCA with this company) and one for BP Solar and all other producers of multicrystalline silicon wafers.

Cooperation with BP Solar was also very close. Representatives from Elkem Solar were allowed to follow the material from filling of the crucible to finished wafers. Representatives from BP Solar also visited Elkem to watch the FFS process and discuss operational difficulties. In the early stage of the cooperation, BP Solar's representatives did clearly express their skepticism regarding material purity. Among their concerns, impurities in silicon could create poor electrical properties or cause cracks in ingots or even evaporate and pollute the furnace walls and roof and cause production failure in several heats. All samples and SoG-Si lots from Elkem were systematically controlled by infrared (IR) measurements for particle detection. Thorough chemical analyses were compared with specification for polysilicon. The impurity content was also compared to the specifications expressed in the early publications of Davis et al. [43] and was found to be far above the critical level. At this early stage, several SoG-Si ingots were returned to Elkem for further studies. Furthermore, competitive cell efficiency had to be demonstrated in order to convince BP Solar.

Elkem Solar researchers had been in contact with University of Konstanz (UKON), Germany since 2002. New samples were sent to SINTEF for ingot production, and cells were made at UKON. The results were presented at PVSEC 14 in Bangkok in 2004 [62] and among other details, the efficiency of Elkem's SoG-Si and commercial polysilicon-based cells were compared.

More than 14% efficiency was obtained on both. In their paper, the authors proudly concluded:

> Commercial level cell efficiency and yield has been demonstrated in PV cells made from SoG-Si produced by Elkem Solar on its proprietary, pilot scale metallurgical refining process. This indicates that the metallurgical refining process developed by

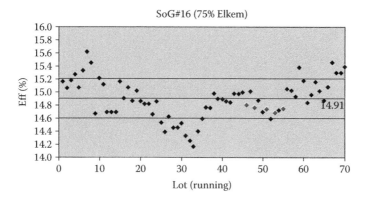

FIGURE 4.5   Test results from BP Solar. Red dots mark the test lots containing 75% Elkem SoG-Si and the blue dots are standard production lots.

Elkem Solar can be developed and up scaled to provide a new, dedicated large volume source of SoG-Si feedstock to the PV industry on a short to medium term horizon. Thorough characterization of the ingots, wafers, and cells produced indicates that:

- Cell efficiency above 14% and diffusion length above 200 µm can be achieved provided resistivity remains above 0.6 Ω · cm.
- Cell efficiency increases with increased resistivity as observed for a single doped material.
- The material responds well to iso-texturation, increasing efficiency by 0.5% absolute. The impact of iso-texturation increases with reduced resistivity [62].

Several tests were also carried out at BP Solar [66] and as seen in Figure 4.5, cell efficiency from test runs followed the standard process variations. Based on these convincing results, a silicon purchase contract between BP Solar and Elkem Solar was signed on September 4, 2006 [65].

4.7.3.2   Cooperation with Q-Cells AG

In 2006, the first contact between Q-Cells and Elkem Solar was established, and in November the same year, a TCA was signed. Q-Cells was a solar cell producer with no internal ingot or wafer production capacities. In the development program for the following years, Elkem supplied samples to both European and Asian wafer producers. These were companies with which Q-Cells had established a wafer supply chain.

At a meeting in October 2006, the first test of ESS (registered trade name of Elkem silicon from March 1, 2004) was planned. A 50/50 mix of polysilicon and ESS should be crystallized and wafered at Pillar in Kiev, Ukraine, and cell production should follow at Q-Cells plant in Thalheim, Germany. From Pillar, the following feedback was returned after solidification of multicrystalline ingot [67]:

- Charge weight of 310 kg. In charge, we used 150 kg (48%) of experimental material.
- We did not notice any abnormality in the growing process.
- Quantity of SiC inclusions on top of ingot is similar to standard ingot.

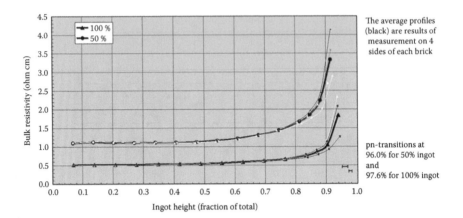

FIGURE 4.6 Resistivity curves for the first two ingots produced by Pillar based on 50% and 100% Elkem Solar Silicon™. (Adapted from Meeting Q-Cells-Elkem, Experiments MG-Si 50%/100%, Internal presentation, Amsterdam, 11.12.2006.)

Also, a 75 kg mono ingot with the same 50/50 mix was produced with acceptable results. The next multicrystalline ingot consisted of 100% ESS and the test procedures and results were similar to the first test (see curve in Figure 4.6).

In November 2006, Q-Cells tested multicrystalline wafers made from 100% and 50% ESS (and 50% poly) with good results. Both test runs used 1500 cells with alkaline and 1500 cells with acidic texture and the same amount of wafers from a reference ingot, grown in the same furnace after the experimental ingots. The efficiency of the test cells showed no difference compared to the reference cells and Q-Cells concluded:

> Elkem-Si has a good potential to be used in large quantities in the production of multicrystalline solar cells.

Q-Cells also referred to Elkem in the yearly report for 2006, and announced the importance of the signed supply contract for their 2010 goal of quadruple production compared to 2006:

> We are convinced that the long-term contract announced at the beginning of February will give Q-Cells a considerable competitive advantage.

The two purchase contracts with BP Solar and Q-Cells, respectively, covered the entire designed volume from Elkem's scheduled industrial plant, and more than satisfied the request by Elkem Solar's board related to sales contracts.

4.8 ASSEMBLING THE PROCESS (2000–2006)

4.8.1 VERIFICATION OF PROCESS TECHNOLOGY FROM RAW MATERIALS TO SoG-Si

The second condition for industrialization was related to the critical steps in the process chain. Each process unit was in principle known from earlier experience in Elkem [68], but to take the process units from a small pilot scale into industrial-sized

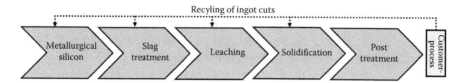

FIGURE 4.7 Elkem Solar process chain. (Adapted from R. Tronstad, 2012. A sustainable product for the solar silicon market, in *Proceedings of the Silicon for the Chemical and Solar Industry XI*, Bergen-Ulvik, Norway, June 25–29, pp. 197–206.)

equipment was a challenge with respect to purity, safety, and productivity. Goals on productivity, cost, purity, energy consumption, EHS (environment, health, safety), and downstream properties were defined for each process step. The five process steps are depicted in Figure 4.7. In the first step, raw materials are specifically selected to produce MG-Si from the furnace with optimal analyses for the subsequent refining steps. In the second process step, slag treatment, boron in the liquid silicon is equilibrated with calcium silicate slag. After the separation of slag and silicon, the liquid silicon composition is adjusted to the leaching ability. The third step, leaching, takes care of removing the main metallic impurities and phosphorus. The fourth step, DS in a crystallizer specially designed for purification, fulfills the elimination of metallic impurities and the reduction of phosphorus. In the fifth and final step, the resulting ingot from step 4 is cut into blocks after cropping (thin layers removed from top, bottom, and sides; these are recycled through one of the former steps depending on their purity).

The process chain is a result of 25–30 years of development work. In the 1970s and 1980s, the idea was to produce pure silicon directly from the furnace. The learning from the project with Dow Corning and Exxon showed that both leaching and DS had to be included. In later projects with MG-Si as starting material, the boron content was too high and slag treatment was introduced as the cheapest and most controllable process for this purpose.

4.8.2   Step 1: Raw Material Mix and Furnace Process

The process steps were more or less concluded already in 2002, and the question of tapped Si quality out of the submerged arc furnace became important to answer. Mapping Elkem's experience and analyses from global MG-Si suppliers, B and P as low as 3–5 ppmw were known. At a meeting on June 27, 2002, in Trondheim, Norway, it was therefore decided to start a subproject whose goal was to optimize the raw material mix to reach tapped Si analyses of maximum 5 ppmw B and the same level for P. At the meeting, Ragnar Tronstad represented the Elkem Solar project while SVP Erik Løkke Øwre and CTO Halvard Tveit represented the silicon division. The subproject got the internal name "Si 5-5." The industrial verification was decided to be run on furnace Nr. 11 at the Fiskaa plant, and Aasgeir Valderhaug was appointed project leader.

After some initial small-scale tests, the test program was scaled up and a set of raw materials was selected. It turned out to be more challenging than expected with a pure charge although the possibility of producing a Si 5-5 quality was finally demonstrated. However, the selected pure charge mix turned out not to be the most economical and productive one, and the B/P goals were accordingly modified in accordance with the more conventional raw material mix.

The furnace process was the first step to be verified in the Elkem Solar process chain [70]. Beside purity, one of the overall goals to the project was to minimize energy consumption through the entire process chain. This included appropriate routines for transportation of liquid Si from the furnace to the slag treatment (step 2).

4.8.3 STEP 2: SLAG TREATMENT

Boron is an element which is removed by DS only to a limited degree. In the early SoG-Si projects, the process strategy was therefore to use raw materials with low B content. By selecting optical quartz qualities, carbon black, and pure binders in agglomerates, it was possible to reach reasonable low B levels, but leaching and DS were needed to get acceptable purity in the produced silicon. This process route was dependent on a stable, low B content in raw materials without any addition of B-containing impurities during handling and transportation from mining to charging on the furnace. From a product quality point of view, it was already obvious from the start-up of the Elkem Solar project that a separate B removal process was needed in order to keep stable electrical properties in the produced silicon. Several technologies have been available, but with Elkem's metallurgical background, slag treatment was selected as the most promising technology. In this process, B removal is a result of an established equilibrium between liquid silicon and a calcium silicate slag. The distribution of B is determined by the L_B coefficient, defined as the ratio between the B concentration in the slag versus that in the silicon. Schei et al. [33], referring to a work of Suzuki et al. [71] who showed that L_B is a function of the CaO/SiO_2 ratio (see Chapter 3, Figure 3.17), repeated the experiments as a part of the internal project and in their own work and confirmed that L_B takes a value close to 2 [72]. This was the starting point for the Elkem Solar project.

In 1999, an important project for the Norwegian solar industry was launched; "From Sand to Solar Cells" was funded by the Norwegian Research Council involving the participation of ScanWafer, Elkem, NTNU (Norwegian University of Science and Technology), and SINTEF [73]. The project established a fundament for education and research on silicon-based solar cell materials at NTNU and SINTEF and strengthened a structured competence buildup at the industrial partners. After a couple of years, the project was split into two subprojects, one open for all parties (FSTS I) and one more confidential project, bilateral between SINTEF/NTNU and Elkem (FSTS II). In FSTS II, Elkem started a study of reactor concepts for the slag treatment. Mathematical models of an electric arc furnace reactor, countercurrent reactor of various designs, semicontinuous ladle reactor, etc. [74] were developed. Laboratory/pilot-scale experiments were carried out to verify the calculations. Consequently, a semicontinuous reactor was selected for industrialization.

The slag treatment experiments also gave feedback on L_B and on metal equilibrium distributions between liquid slag and liquid silicon. Measured B content in slag and silicon indicated a higher L_B value than earlier published data. A separate study

FIGURE 4.8   Pilot operation at Elkem Solar plant.

was needed to understand the reason for such discrepancies. A systematic analytical error was detected as the reason for large deviations in L_B. After system correction, L_B was calculated with higher accuracy to 3 ± 0.5 [75]. The accuracy was later improved when more data became available.

Slag for B removal was produced in a separate process step and added as liquid to the slag treatment reactor. It was important to get raw materials with low content of impurities and especially low boron. Thanks to Elkem's globally widespread presence and established network, it was possible to secure long-term supply contracts on stable raw material quality.

The development of furnace for slag production built on Elkem's earlier experience as a supplier of smelting equipment for different slag processes. Process experience was gained during pilot production as all SoG-Si produced for testing at institutes and by future customers had to be slag treated. Modifications of slag composition, process temperature, lining systems, etc. were established in this period. The slag treatment process was finally validated in 2006 [76] after having produced, during 2005, more than 15 MT of slag-treated Si in the pilot plant (see Figure 4.8) with convincing results on B removal, product stability, process robustness, etc. The test period had also emphasized the importance of reliable analytical methods and stable raw material quality with low B content. These factors retained accrued attention during further testing in the pilot plant.

4.8.4   STEP 3: HYDROMETALLURGICAL TREATMENT: LEACHING

The Silgrain process at Elkem Bremanger was invented more than 40 years ago and purified silicon (Silgrain) from this plant is still an important part of Elkem's product

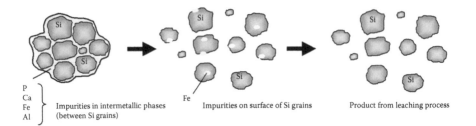

FIGURE 4.9   The leaching process.

portfolio. For some years, Silgrain covered more than 90% of the global MG-Si supply to polysilicon production (Siemens process). The process idea is based on removing the impurity-rich phases surrounding each grain of silicon and thereby disintegrating the lumpy and impure silicon alloy into small particles as illustrated in Figure 4.9. Most of the impurities accumulate between the silicon grains during solidification. Removing this layer is therefore a purification process.

Silicon from the slag treatment contained both metallic impurities and high amounts of phosphorus. The established Silgrain process did not satisfy the SoG-Si demand. Modifications were introduced in several steps, first in the Exxon period [38,39], then in 2001 [77], and also later in the Elkem Solar project period.

Most of the leaching development in the Elkem Solar project has taken place at Elkem's research center in Kristiansand, Norway. A bench-scale laboratory and a 30 kg pilot plant for leaching experiments gave valuable input to process design and intermediate products. Purified silicon from the 30 kg pilot plant covered the need for test material in the DS process and sampling for solar cell validation at the University of Konstanz/UKON (after 2005 at ISC-Konstanz).

Final validation of the leaching process took place at the Bremanger plant in existing pilot equipment for Silgrain process development ("SIMO" building). The first tests were performed during the autumn of 2004 with a capacity of 3 MT per batch. After some initial campaigns, the pilot was redesigned to a semi-continuous process based on engineering input from the existing industrial plant. This took place in 2005 and testing could start in week 42. Results from this campaign and the following tests in Q1 2006 confirmed the results from bench- and 30-kg-scale tests. The product quality was confirmed, and except for some technical issues with filtering, all process checkpoints were successfully passed [78].

4.8.5   STEP 4: DIRECTIONAL SOLIDIFICATION: A NEW COMPETENCE AREA FOR ELKEM

Experience from earlier SoG-Si projects at Elkem had demonstrated the need for a DS process as the final refining step. Cooperation with AstroPower had given access to their "recycling machine"—an industrial version of the CUDS technology. The development work was, however, far behind schedule and key scientists involved at AstroPower had quit their positions [79]. Therefore, Elkem Solar decided to start its own research activity on DS.

There were several suppliers of commercial Bridgman equipment, but both *capex* and *opex* seemed too high for Elkem Solar's purpose. It was therefore decided to start an internal project and gain competence on DS, through small-scale experiments and product evaluation studies. The first experiments took place in November 2002 and clearly demonstrated that DS of silicon, which had passed slag treatment and leaching, was a way to pursue. The simple equipment used in this test was given the name FFS—furnace for solidification—and consisted in principle of an induction furnace and a crucible [80].

In order to verify the solidification technology in a close-to-industrial-scale pilot, an intermediate-size unit—designed on the same principles as the pilot—was established. Troubleshooting in such a unit should give valuable input to design the industrial unit and, hopefully, save development time. The small unit got the name FFS1 and had an ingot size of 70 kg. Measurements during solidification and study of ingots from FFS1 gave valuable input to solidification models under development at SINTEF/IFE and helped optimize the process operation strategy. Produced ingots (see Figure 4.10) were cut into bricks, surface cleaned, and sent to cooperation partners for testing in ingot production, wafering, and solar cell lines. In 2005, the design of FFS2 (close to industrial-scale unit) gained speed. With an ingot size of 300 kg, the "close-to-industrial-scale" demand was satisfied. The design of FFS2 became rather complex with both a preheating and a cooling chamber, respectively, in front of and after the solidification position. These two units were soon eliminated and the verification could be performed in an enlarged version of FFS1. When moving from 70 to 300 kg ingot, scaling problems occurred and delayed the verification by 6 months.

In the meantime, discussions had started with the German furnaces supplier ALD on the design of customized equipment based on the FFS concept. It was also decided to build a new solidification pilot (named multi mold crystallizer—MMC), where a copy of the industrial unit could undergo careful examination. An old building at

FIGURE 4.10   Section of the first ingot produced by Elkem Solar.

Elkem Carbon—close to the research center—became the new solidification pilot. It took 8 months from the first growth in December 2007 to the final report [81] certifying the successful production of ingots in a prototype furnace for industrial DS.

4.8.6 STEP 5: POST-GROWTH TREATMENT

The posttreatment step consists of cutting the ingots into bricks, removing top, bottom, and side layers, cleaning surfaces, recovering Si fines from waste water, controlling the quality of bricks, and packing. When silicon from the leaching plant enters the solidification unit, it contains both particles and metallic impurities. Particles will normally settle at the bottom of the crucible during melting; some particles will be generated during solidification and transported to the crucible sides. Both particles and dissolved impurities will accumulate in the last residual melt. The bottom, top, and sides therefore have to be removed and recirculated to previous steps in the process chain. As the value significantly increases along the process chain, an optimal cutting process needs a high yield and should keep the product clean. Thickness optimization of bottom, top, and side cuts had a major influence on the yield and also on selecting the cutting blade dimensions. It was decided that the weight of each brick should not exceed 10 kg and the dimensions should be within certain limits to satisfy a standardized filling of crucibles in the customer's process.

In 2005, Elkem Solar built a separate pilot for the post-taphole activities in a rented building near the research center where tools and procedures for cutting and surface cleaning could be worked out in close cooperation with suppliers. This development further influenced the operational conditions to the solidification unit as ingots were prone to crack during cutting because of accumulated stress and strain in the material. This undesired phenomenon was more pronounced with increased ingot size and therefore was most critical in the MMC unit (350 kg ingot).

A mathematical model of DS of Si had been under development in cooperation with IFE/SINTEF for analyzing segregation, solid/liquid interface, particle movements, heat distribution, etc. The crack problems triggered the development of a new algorithm for the calculation of stress and strain in the ingot. This turned out to be an important tool for adjusting the control parameters to the solidification process, and thereby limiting the problems of cracking during cutting.

Verification tests of the posttreatment step focused mainly on the cutting of ingots. Owing to a constrained schedule, the report had to be completed by the first half of 2006 [82] and for that reason only ingots from FFS2 heats were included in the test program.

4.9 PLANT BUILDING, ENTERING INTO THE COMMERCIAL PHASE, NEW OWNER, AND NEW MARKET CHALLENGES (2006–2015)

4.9.1 PARALLEL PILOTING AND RAISING OF PRODUCTION PLANT

A 2004 document [83] forecast a pilot production at Elkem Solar starting with MG-Si from the silicon plant at Fiskaa, Kristiansand, continuing with slag production and slag refining in the pilot plant at the research center, leaching at the SIMO plant at the Bremanger plant, and finalizing with crystallization and posttreatment at the pilot plant.

A capacity of 100 MT/year could be reached without large investments, which would allow BP Solar to conduct long-term testing of the product in their lines and strengthen the joint business case. This plan was, unfortunately, not fulfilled. One of the reasons was the closure of the MG-Si plant at Fiskaa/Kristiansand in the beginning of 2005. During the following year, new industrial plans were worked out, forecasting capacities of both 500 and 2500 MT/year before a new scope document [84] suggested a capacity of 5000 MT/year. This latter capacity would balance the capacity of either furnace 11 or 12 at the idle Fiskaa plant and represented an improved cost scenario. The corporate management received the investment application in September 2006, and a positive "go" came from the Orkla group, at that time the majority owner, 1 month later. The board had also decided that the commercial plant would be built as one unit in Kristiansand and not in Bremanger, which was an alternative discussed in the application. The proximity to Elkem's corporate research center in Kristiansand (see Figure 4.11) had a major influence on this decision, which disregarded the advantage of a joint leaching infrastructure and competence at the Bremanger plant.

The solar industry floated on a wave of optimism in 2006 with a yearly market growth of more than 30% and a price level of $40—$50 per kg Si. However, forecasts predicted falling prices in the years to come and it became important for the

FIGURE 4.11   Elkem in Kristiansand, Norway.

board to build the plant as fast as possible. With two sales contracts in hand and some incomplete process verification, the building of the first Elkem Solar plant started [85]. The project cost was estimated to 2.7 billion NOK and the plant was to be ready for commissioning on October 29, 2008.

The project faced big challenges:

- Technology development and design in parallel with project execution
- Part of the plant placed in a brownfield area
- No operational experience on several process steps
- Operating organization in parallel with project execution

A risk analysis performed by a consulting company also pinpointed the technology risk and addressed the stressed market situation at equipment suppliers and consultants, which could delay the project and increase cost [86]. Consequently, the project management implemented Orkla's framework for risk management. This systematic tool contributed to a successful implementation of equipment and process.

An internal revision team addressed the same items and in addition considered the project execution risk to be high. Their report concluded that Elkem did not have systems, competence, or experience to run projects of this size. Therefore, Elkem Solar signed Engineering Procurement Construction Management (EPCM) contracts with two external companies, Fluor and Hydro Production Partner (HPP—sold to Bilfinger Berger Industrial Services AG in 2008), to reduce this risk.

The project obtained great success on environment, health, safety (EHS). No accidents causing an employee being away from work, or unable to perform normal work duties, were reported. This is the best result ever for a land-based industry project in Norway. However, the project did not meet scheduled time and cost, and the evaluation report [87] uncovered a number of reasons. The quality of existing buildings did not satisfy the official regulations and unscheduled refurbishment work increased the cost considerably. The same applied for ground work. Installation of a huge number of concrete pillars was necessary for the stability of the new building hosting MMC units and posttreatment. Another factor that influenced both time and cost had to do with the change from 2500 MT/year to 5000 MT/year SoG-Si. All pre-engineering works done in 2005 were based on a 2500 MT/year capacity and were already executed before finishing process verification. The change to 5000 MT/year SoG-Si was decided in the spring of 2006 and that initiated a new localization debate only half a year before demolition and construction work started. Despite the delay of 7 months and an increased cost of 1.7 billion NOK, the process units were received and installed close to budget estimates.

In the autumn of 2006, the board was aware of several competing ongoing SoG-Si initiatives:

- Fluidized bed technology was proven at MEMC and under implementation at REC.
- The JFE holding in Japan produced approximately 800 MT/year SoG-Si with their upgraded metallurgical process. This was captively consumed by JFE.
- Dow Corning tested their UMG-Si process alternative in Brazil.

- Timminco patented an SoG-Si process also based on UMG-Si.
- 6N Silicon (a Canadian startup company) promoted a new UMG-Si based on dissolving impurities in aluminum.
- The Solsilc concept, a cooperation between Fesil, SINTEF, and SunErgy, explored the route of pure raw materials.
- Several others.

The technical team at Elkem Solar knew several of these processes from both confidential evaluation meetings and from open literature studies. In some cases, common development and even joint venture alternatives had been discussed. At the end of the meeting, Elkem choose to improve its own technology and take the pilot results to industrial-scale production.

When the MG-Si plant closed at Fiskaa in 2005, operators from all parts of the production were offered new jobs in Elkem Solar. Operating a silicon furnace needs years of training and starting up the Elkem Solar process without this expertise would be risky. The operators took part in the piloting and learned a lot through troubleshooting and close discussions with the specialists in charge of the different production technologies. Elkem also started an "Elkem Solar School" for both operators and technical staff to introduce the theoretical background for the critical items. All employees at Elkem Solar received this education. When commissioning started in the summer of 2009, all personnel were extremely well trained and could from day one excel in their contribution.

Shortly after production started, the project group presented a concept study on the next plant. The new plant would have double capacity and be a greenfield design. Owing to the financial crisis of 2008 and difficult times for the PV industry in the following years, this study has, at time of writing (2015), not materialized.

In parallel with process development and engineering, important achievements in the chemical analysis of dopant elements and metals followed the increased demands for higher Si yield in downstream processes and improved resistivity control in produced wafers. New equipment and procedures have lowered the detection limits from 5 ppmw for B and P to 0.15 and 0.4, respectively. The new methods (Table 4.4) are part of Elkem Solar's quality control for both process and product.

4.9.2   INDUSTRIAL PRODUCTION OF SoG-Si IN KRISTIANSAND

Prince Haakon of Norway officially opened the plant on August 21, 2009. Business acquaintances from all over the world were present. Speeches and entertainment made the day unforgettable for those attending the event. More than 30 years of R&D had finally created a new industry in Norway. The goals had been met [88] on energy consumption and product quality, and all estimates indicated a low production cost for ESS.

Statistics from other industry projects indicated a ramp-up time to full capacity in the range of 2–3 years. The capacity had been upgraded from 5000 MT/year to 6000 MT/year during the construction period, but the real production was far behind this level during the first year with a monthly production below 150 MT. The process steps where Elkem had long industrial experience worked well, but in steps 4 and

TABLE 4.4
Development of Chemical Analyses at Elkem Solar

Period	Detection Limit for B (ppmw)	Detection Limit for P (ppmw)	Comments
Before 1985			Simple wet chemical methods, ca. 5 ppmw B, 2.5 ppmw for P
1985–1998	5		First ICP-OES in 1995
1998–2004	1		New ICP-OES in 1998, B and P from the same sample
1998–2007		2.5	New ICP-OES in 1998, B and P from the same sample
2007–current		2	Improved method for B
2003		0.5	New method for metals and P with ICP-OS
2004	0.2		New method with ICP-MS for B
2010		0.4	Improved method for bigger samples
2012	0.15		Improved ICP-MS method for B
2014			Detection of a new B species, which results in biased ICP-OS method [89]

ICP-OS = inductively coupled plasma optical emission spectrometry.
ICP-MS = inductively coupled plasma mass spectrometry.

5 (DS and cutting), cracking problems with both ingots and crucibles caused low silicon yield and long stoppages. The corporate R&D center—Elkem Research—became a part of Elkem Solar Research for a period of 2 years to help solve the problems. Systematic problem-solving gave large improvements in the last half of 2010 with demonstration of 300 MT/month. Improvements continued through 2011 with increased capacity and technical performance. Unfortunately, this was the year when China raised the production of SoG-Si and most of the manufacturers in the solar industry had problems. The global production capacity of SoG-Si, wafers, cells, and modules increased incredibly faster than the demand; prices of SoG-Si fell from $200 to $40/kg and similar price movements were seen along the entire value chain. Q-Cells was one of Elkem Solar's customers who suffered most. At the end of the year, the negative result had accumulated to €846 million, and on April 4, 2012, Q-Cells stopped production. Before Elkem Solar started production, the supply contracts to the two biggest customers had covered almost the entire capacity of the new plant. With Q-Cells out of business, a significant volume was made available, and under the global financial crisis, it turned out to be difficult to compensate the contract loss. In the last part of 2011, the factory was forced to produce at a lower capacity. Some new contracts had been signed, but also the new companies had financial problems and on September 19, 2012, Elkem Solar temporarily stopped production.

In January 2011, Orkla signed an agreement with China National Bluestar Co, Ltd (Bluestar) on the sale and purchase of the entire Elkem group. Bluestar is an international company with production sites in most continents. In Europe, they already had a 3-year ownership of Bluestar Silicones International (BSI); MG-Si is the most important raw material for this company and its counterpart in China. With Elkem in Bluestar's portfolio of subsidiaries, the supply of MG-Si was secured. This takeover turned out to be very important also for Elkem Solar, first of all due to the new owner's long-term vision on strategic developments. During a standstill from September 2012 to January 2014, only a few people from the plant were laid off. Most of the employees got meaningful engagements at Elkem Research as well as other Elkem and Bluestar plants. New development projects engaged the technical staff and when the plant resumed production on January 28, 2014, a motivated workforce returned back to the plant.

4.9.3 Proving the Product Capability in the Field

Looking back to the period 2012–2014, most of the difficulties affecting the Western PV companies found an easy explanation in the competition with the massive offer of cheap Chinese products. For Elkem Solar, the situation turned out to be more complex. ESS contained both acceptors (B) and donors (P), so-called compensated materials, which was not well known among wafer producers. By controlling the amount and ratio of B and P, high-quality wafers with predictable properties could be made. Resulting from a metallurgical production process, the product was classified as UMG-Si material (upgraded metallurgical-grade silicon). Purified silicon from Timminco Solar, a subsidiary of Bécancour Silicon Inc., a Canadian company, had the same classification. At a Photon SSC conference in 2008 [90], Timminco presented chemical composition of their UMG-Si quality and showed questionable variations in the doping elements. When the company ceased production in 2009 (stopped definitively in 2010), customer claims regarding the product quality were publicly known. After this adversity for the Canadian company [91], Elkem Solar also suffered from the UMG-Si reputation and struggled to convince existing and new customers about the ESS quality. Thanks to long-term research programs and continuous development of each process unit, Elkem Solar built up knowledge and documentation on ESS quality compared to standard solar-grade polysilicon. These studies documented efficiency performance and degradation over time at par with polysilicon. The most sensational results came from long-term scientifically controlled tests in India (see later).

Since the start of the Elkem Solar project, close cooperation with both research institutes and potential customers have been an important part of the strategy. The basic understanding of each step in the value chain needed years of research together with Norwegian universities and institutes while studies of ESS performance in downstream processes also included several internationally recognized research groups and institutes. The data, models, and new knowledge developed in these programs documented the product attributes and the process capabilities, widened the professional network, improved the knowledge level, and had a major influence on troubleshooting skills during the entire project period. ESS represented an unknown feedstock to most of ingot/wafer producers and good documentation on quality was critical to get access to the market. Since ESS contained both B and P, a new model for net-doping calculation guided the customers in selecting the right mix of Si feedstock [92], and tests of ESS in new cell concepts proved the product capability over a broad range of cell technologies. A number of results obtained with industrial tests, silicon quality, specification, degradation, and capability tests in advanced cell concepts were published together with cooperation partners [93–111].

Cooperation with existing and potential customers generated a lot of information on the use of ESS in different multicrystalline DS furnaces. Wafers from Bridgman-type and continuous solidification furnaces have been followed all the way to solar cells and some to modules. One of the most interesting results connects to module tests near the equator. At Hyderabad in India, Elkem Solar installed modules made of ESS and reference modules of polysilicon in 2009 [112]. Both test and reference material went through equal treatment at the same producers all the way to modules, that is, the same ingot furnace produced ingots of polysilicon and ESS at the same time period, wafers cut in the same saw, modules made by the same producer, etc. After installation and 1.5 years of operation, ESS modules and reference modules changed place in order to eliminate any effect of shadows and other position-sensitive parameters. The modules have now (in 2015) been in operation for 6 years and the yearly energy production for all years has been more than 2% higher for the ESS-based modules than for the reference modules. The November 2015 results show more than 5% higher energy production (Figure 4.12). The same tendency is observed from test sites at Ishinomaki, Japan, and Australia [113].

Another competitive advantage is the low energy consumption in ESS production. In the first life cycle assessment (LCA), data input originated from pilot operation. This analysis covered the entire value chain from mine to final product and in comparison with polysilicon produced by the Siemens process, only 25% of the energy consumption was needed [114].

A new LCA analysis in 2012 verified the consumption figures, but this time for the industrial plant [115–117]. Figure 4.13 shows a plot of energy payback time (EPBT) for ESS compared to silicon produced by a modified Siemens process in the EU and China. "Modified Siemens" in this context means Siemens with TCS recycling

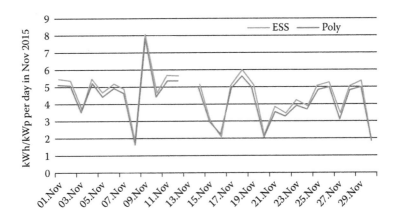

FIGURE 4.12   Power production (kW/kWp) per day at the Hyderabad PV plant in India for ESS and reference modules (November 2015).

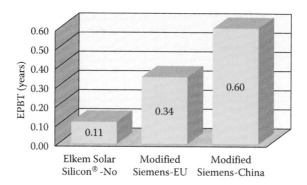

FIGURE 4.13 Energy payback time for Elkem Solar and modified Siemens in EU and China.

and "Modified Siemens-EU" means "Best Case Siemens" with all electricity from hydropower and all heat from co-generation power plants. With this documentation in hand, Elkem Solar was the first SoG-Si producer that was awarded green product status in Japan and obtained a similar status in France.

4.9.4 Status and Direction of Development for Elkem Solar: 2015

In 2006, Elkem Solar and Blitz Solar (later Berlin Solar/Calisolar/Silicor Materials) started a technical cooperation program. Blitz Solar was a newcomer, but had very competent people with excellent background in solidification and cell processing. The laboratory in Berlin is equipped with Bridgman furnaces and dedicated instruments used for studies of process details relevant to solar cell production. One of the ideas that came up at an early stage of the common program had to do with simplification of the value chain. DS was the most expensive process step in the Elkem Solar process chain, and the first step in the customer process. Was it possible to eliminate a solidification step and/or substitute it with a cheaper refining technology? The project was named SALO, an acronym for "Slag and Leaching Only." In Figure 4.14, the first ingot in the SALO project is shown. Slag accumulation on top of the crystallized silicon and disappointing chemical analyses did not satisfy the specification. However, the search of simplifying the process chain continued at Elkem Solar and two projects later Elkem Solar was able to publish solar cell results from a simplified process, with efficiency equaling standard polysilicon-based cells [118–120]. The simplified process is ready for implementation in the future expansion of Elkem Solar and will contribute to significant competitive advantages when in operation.

4.9.5 Building a New Integrated PV Company (2014–2015)

In February 2014, production resumed at Elkem Solar. The market demand and price had slightly improved compared to 2012. However, supply was still abundant from the legacy producers of polysilicon by the Siemens process. Tariff and antidumping

FIGURE 4.14   The first SALO ingot.

duties imposed on products (on Chinese modules into the United States and on U.S. polysilicon into China) contributed to complicate global market dynamics. Elkem Solar's position as an independent producer of a unique but different product (dopant compensated silicon) with a modest capacity (5–6000 MT/year) was extremely vulnerable to any change in the market. Cooperation with a partner downstream in the value chain was an obvious strategic direction to explore. In December 2014, Bluestar, Elkem, and REC Solar announced the intention of Elkem Solar to acquire REC Solar and thereby establish an integrated PV business under the umbrella of Bluestar. The transaction was finalized at the end of August 2015, but the integration work between the silicon producer in Norway and the wafer-to-module producer in Singapore had already started in the very first days of the year. Furthermore, in August 2015, Elkem announced that it would acquire the idle ingoting plant of REC at Herøya (Norway) from the bankruptcy estate manager. After upgrading the existing furnaces to today's technology, the ingot plant would have a capacity equivalent to 3500 MT/year ESS consumption. Little is known about the decision process that led to these remarkable achievements. The events are too recent to be treated as historical studies. However, it seems obvious that the common history of both groups as shown in this chapter, and certainly also the personal relationship between the key executives, active from time to time in one or the other entity, played a significant role in the march to unity. On the technical side, it is known that intense evaluation of ESS in REC Solar modules in the course of 2014 led to convincing results, encouraging both parties to accelerate their cooperation.

As an integrated company from silicon to modules, the new Elkem–REC Solar entity feels better positioned in the global market. As of November 2015, the production capacity at Elkem Solar has crept to 6300 MT/year, the stock is empty, and no material is available for new customers. REC Solar, which enjoys a strong demand from the U.S. market, is also sold out and looking externally for cells to supplement its internal capacity. However, long-term profitability as well as huge capital requirements to keep investing in capacity remain serious challenges for the management and the owner.

4.10   CONCLUDING REMARKS

After 40 years of development, Elkem and other Norwegian companies have established a solid foothold in the PV industry. This is true not only for the newly launched Elkem–REC Solar group but also for Scatec Solar, an Oslo-listed company, building and operating solar parks throughout the world as well as for NorSun and Norwegian Crystals, two producers of single crystals who remarkably survived or resurrected from the financial crisis (2008–2009) and the following big shakeout in the PV industry. To stay in the race, however, requires significant renewed effort and investment. With 6500 MT/year capacity of SoG-Si, even expanded to 7500 MT/year in 2016, Elkem Solar has a modest capacity compared to the 60,000 MT/year at each of the two largest producers and the current global output capacity of around 300,000 MT/year. The new combined Elkem–REC Solar group will have to invest not only in feedstock capacity, but in all steps along the value chain. This is highly capital intensive and will need prioritization of capital allocation at the right times.

The earlier hypothesis of vertical integration (already formulated at Elkem during the 1980s) is re-confirmed with full strength. An obvious and necessary issue today is to clarify what to do in Norway/Europe and in Asia. Thanks to available hydropower and access to highly qualified technical competency, Norway seems competitive upstream, whereas more labor-intensive operations such as cell and module manufacturing are more appropriate at locations where labor is more affordable or closeness to large end-user markets is a consideration.

Our historical review shows that the main steps of the current Elkem Solar process were identified and known already in 1985. We believe that a steady and continuous effort—without the frequent "stop-and-go"—would have changed the course of history. This is true also for the strategic downstream development and integration. Actively supporting Crystalox to develop its 1985 ingot and wafer plan and keeping it inside the Elkem group (versus MBO in 1994) would at an early stage have positioned the group as an integrated PV leader. This is of course pure hypothetical speculation. History never follows mechanically linear paths. History is always the result of human interactions and the resolution of large and small external as well as internal conflicts. The reason why it took altogether 40 years instead of 15 or 20 years and cost twice or five times more than necessary lies not only in the failing market demand or in the limited resources but also in the decision process in the company itself under the strong influence of the prevailing business and management culture. It was not negative technical outcomes that led the management to adjourn or suspend the solar cell project in Elkem (1985 and 1994), but the lack of knowledge of and insight into the long-term trends.

Short-term thinking of 5–10 years is the normal time horizon for most decision makers in Norwegian companies. A longer perspective seems more the rule for the new Chinese owner. After it took over Elkem, Bluestar ChemChina has encouraged further technology development and maintained equipment in perfect shape also during the long period of almost 2 years (2012–2013) when production at Elkem Solar was put on hold. Shortly after production was resumed, Bluestar announced (2014) the acquisition of REC Solar in Singapore and the practical merger of these activities with Elkem Solar, thus reshaping a new powerful integrated PV player from silicon to module. The later decision (September 2015) to reopen an REC plant in Norway for the production of ingots confirms the perception that Norway is an appropriate place for certain PV manufacturing activities. This is the type of action and confidence in the long-term future that was lacking in Norway when the young and fragile PV industry ran into severe difficulties as a consequence of the financial crisis starting in 2008. Why did the Norwegian government in 1991 rescue the ferroalloys producer Elkem, but did not make a single attempt 20 years later to preserve the high-technology workplaces at REC and its service providers?

The whole Norwegian solar adventure would not have been possible without the strong technology background and the technical ability preexisting in the country [121]. It was not a pure coincidence that led companies with solar ambitions to Elkem (Research and Bremanger). There they found metallurgists, researchers, and experts able to solve the raw material challenge. As opposed to the upper-level management, the technical and technology staff was remarkably stable. These social groups were those who could take care of the good ideas and keep alive the collective technology memory of the company through all up- and downturns. Elkem Research's archive, which we have consulted, contains a rich collection of detailed notes and reports exhibiting a deep knowledge and understanding of our own and competing technologies. Several among Elkem's competitors made serious attempts to develop a metallurgical route to SoG-Si. None came as far as Elkem. Elkem was probably the one with the largest diversified technology background, a long history, and unique capability in R&D (since 1915). This is an invaluable asset for the new Elkem–REC Solar to bring the further work toward higher objectives.

Such a strong technology background was not the attribute of the young company REC when it was founded. Production of ingots, wafers, cells, and modules as first practiced in Scandinavia and then transferred to Singapore were at best weakly proprietary. Equipment and processes were originally provided by commercial suppliers, who rapidly transferred to their customers any improvement achieved or observed at other places. This quick technology dissemination responded to the fast growth of the PV market and made it possible. It profited REC, and also its competitors. REC gained access to a strong technology portfolio when acquiring ASiMI, a successor of Union Carbide Chemicals for polysilicon and silane. The portfolio comprised unique technologies on which REC could develop its proprietary FBR polysilicon process, making a strong differentiation with its competitors. Splitting REC into a silicon company (REC Silicon in the United States) and a solar company (REC Solar in Singapore producing wafers, cells, and modules) was later a financially motivated operation to crystallize the value of and secure sufficient funding for the two businesses. This operation took away the downstream integration from the former and a unique differentiation from the latter. What made REC a success in the beginning was therefore not a particular or original technology platform. The success came from the spirit of entrepreneurship and the strong willingness that the founders demonstrated in the buildup period.

Now the circle seems closed. The plan of an integrated solar company, which was a goal for Elkem already in the middle of the 1980s, has come to a lucky conclusion with one strong pillar in Norway and another in Asia (Singapore) and the whole world as a marketplace. The history shows that the new group has a strong and solid platform to build on, but also that it should learn from past good and bad experiences to face the future and its many challenges.

ACKNOWLEDGMENTS

The authors are grateful to Elkem Technology for providing access to its archive. Special thanks to Iain Dorrity for the personal communication on Crystalox's history, and to Gunnar Halvorsen and Gøran Bye for reviewing this chapter and for their advice.

REFERENCES

1. K. Sogner, 2003. Elkem gjennom 100 år, in *Skaperkraft 1904–2004*, Mesel Forlag, Oslo.
2. B. Ceccaroli and O. Lohne, 2011. Solar grade silicon feedstock, in A. Luque, and S. Hegedus (Eds.), *Handbook of Photovoltaic Science and Engineering*, 2nd edition, Wiley, pp. 153–204.
3. B. Ceccaroli and S. Pizzini, 2012. Processes, in S. Pizzini (Ed.), *Advanced Silicon Materials for Photovoltaic Applications*, Wiley, pp. 21–78.
4. G. Bye and B. Ceccaroli, 2014. Solar grade silicon: Technology status and industry trends, *Solar Energy Material & Solar Cells*, Elsevier, 130, 634–646.
5. V. D. Dosaj, L. P. Hunt, and A. Schei, 1978. High-purity silicon for solar cell applications, *JOM*, 30, 8–13.
6. Letter from Dow Corning to Elkem, 07.02.1975 (internal).
7. Internal report Elkem, F291/75.
8. Letters, Elkem to and from Dow Corning, June 1975.
9. Internal Memo Elkem, 31.03.1976.
10. A. Schei, Starter utarbeidelse av revidert tilbud—100 kW, *Internal Memo Elkem, 20.04.1976.*
11. H. H. Aas and J. A. Kolflaath, 1974. Process for refining technical grade silicon and ferrosilicon by continuous extraction, U.S. Patent 3,809,548.
12. H. H. Aas, 1971. The Silgrain process: Silicon metal from 90% ferrosilicon, *TMS Paper Selection*, Nr. A71-47, The Metallurgical Society of AIME, 653.
13. A. Schei, 1975. Ovn for Si til silikoner og rent Si til elektronikk og solceller, *Internal Memo Elkem, 25.04.1975.*
14. K. Larsen, 1976. Innledende forsøk med rensede råmaterialer, *Internal Report F/FV 81/1976.*
15. Renset charcoal har dårligere reaktivitet, *SINTEF Report STF34 F76031 1976.*
16. Forsøk med renset trekull, *Internal Report Elkem F/FV 79/1977.*
17. K. Larsen and A. Schei, 1977. Trekull renset ved 1700, 2000 og 2000°C/vacuum, *Internal Report F217 1977.*

18. Letter from Dow Corning to Elkem, 18.05.1976.
19. J. P. Hunt et al. 1978. Solar Silicon via the Dow Corning process, *Quarterly Report No. 7, DOE/JPL/954559-5*, April.
20. A. Schei, 1978. Første smelting I 100 kW ovn hos Dow, *Internal Report F47-1978*.
21. A. Schei, 1979. Valg av rene foringsmaterialer—SiO_2 valgt, *Internal Report F65-1979*.
22. L. P. Hunt and V. D. Dosaj, 1979. Final Report. Solar Silicon via Dow Corning Process for the Silicon Material Task, Report No. DOE/JPL 954559-78/7, JPL Publication, 82–79.
23. Th. Pedersen, 1979. Rensilisium for solceller, *Memo, 04.10.1979*.
24. A. Schei, 1980. Samarbeid med Exxon om ekstra rent silisium, *Memo, 03.03.1980*.
25. A. Schei, 1980. De første oppgavene i samarbeidet Exxon-ES, *Memo, 29.02.1980*.
26. Elkem/Exxon program on SAR silicon for photovoltaics, *Memo, 11.01.1980*.
27. A. Schei, 1980. Samarbeid Exxon-ES, *Memo, 08.05.1980*.
28. A. Schei, 1980. Søknad om bevilgning til Solcelleprosjekt, *Memo, 15.07.1980*.
29. Ingeniørforlaget 10 år, www.ioea.org/inis/collection/NCLCollectionStore/.../11514764.pdf, 1978.
30. H. Storegraven, 1980. Ren S-Exxon-muligheter og risiko, *Memo, 08.04.1980*.
31. A. Schei, 1981. Exxon/Elkem joint R&D program on low cost silicon for solar cells, *Elkem Report No. 2 to the Steering Committee/Internal Report F86/81a*.
32. K. Larsen, 1983. Elkem–Exxon program. Startforsøk Nr 1-16, *Internal Report F38/1983*.
33. K. Larsen and A. Schei, 1984. Arc furnace smelting of silicon for solar cells, *Internal Report F16/1984*.
34. A. Schei, 1983. Exxon–Elkem joint R&D program on low-cost silicon for solar cells, *Elkem Report No. 10 to Steering Committee/Internal Report F79/83a*.
35. A. Schei, 1983. Exxon–Elkem joint R&D program on low-cost silicon for solar cells. *Elkem Report No. 9 to the Steering Committee, Internal Report F21/83a*.
36. A. Schei. 1980. Exxon–Elkem joint R&D program on low cost silicon for solar cells, *Elkem Report No. 1 to Steering Committee, Internal Report F30/81a*.
37. A. Schei, J. Kr. Tuseth, and H. Tveit, 1988. *High Silicon Alloys*, Tapir Forlag, Trondheim, ISBN 82-519-1317-9.
38. G. Halvorsen, 1985. Method for production of pure silicon, U.S. Patent 4,539,194, filed 06.02.1984 (approved US, 03.09.1985).
39. G. Halvorsen and A. Schei, 1984. The removal of carbon from silicon, *Internal Report F35/1984*.
40. A. Schei, 1983. Removal of boron from silicon by treatment with the gaseous mixture H_2-H_2O, *Internal Report F114/1983*.
41. J. P. Dismukes, 1983. Final Exxon report to the steering committee for the quarter October 1, 1982 to December 31, 1982.
42. J. K. Tuseth and O. Raaness, 1976. Reactivity of reduction materials for the production of silicon, silicon rich ferro-alloys and silicon carbide, *Proceedings of the Electric Furnace Conference*, 34, St Louis, USA, 101–107.
43. J. R. Davis Jr., A. Rohatgi, R. H. Hopkins, P. D. Blais, P. Rai-Choudhury, J. R. McCormick, and H. C. Mollenkopf, 1980. Impurities in silicon solar cells, *IEEE Transactions on Electron Devices*, 27, 677–687.
44. J. P. Dismukes, 1984. Solar cell performance assessments of Elkem–Exxon silicon, *Internal Report F114/1984*.
45. T. Ulset, 1997. Summary, Solar Silicon Seminar, *Memo, June 4–5*.
46. Letter from IMC (International Minerals & Chemical Corporation), 07.02.1985.
47. T. Kjelland and R. Tronstad, 1986. Si-smelteforsøk I 500 kW ovn. Innledende forsøk for smelting av solcellesilisium, *Internal Report F52/1986*.
48. A. Schei, 1993. Equipment for Eurosolare program, *Memo, 3.05.1993*.

49. R. K. Brenneman, A. Schei, R. Kaiser, and Y. Lee, 1992. Issues in solar-grade silicon feedstock development, *Proceedings of the 11th E.C. Photovoltaic Solar Energy Conference*, Montreux, Switzerland, pp. 412–415.
50. A. Schei, 1993. *Internal Report Elkem F121.*
51. J. D. Esdaile, G. W. Walters, and J. M. Floyd, 1972. Australian Patent Application 48.570, November 3.
52. J. D. Esdaile and G. W Walters, 1975. Continuous refining of metals, Australian Patent Application 85.129, September 19.
53. J. D. Esdaile, A. B. Whitehead, G. W. Walters, and W. T. Denholm, 1977. Reflux refining of metals, Australian Patent Application 26.858, May 16.
54. J. D. Levine, G. B. Hotchkiss, and M. D. Hammerbacher, 1991. Basic properties of the Spheral Solar™ cell, *Proceedings of 22nd IEEE PVSC*, Las Vegas, USA, vol. 2, pp. 1045–1048.
55. McKinsey & Company, 25.20.25.5 by 2005, Executive summary, 15.05.2001.
56. B. Ceccaroli, 1999. Solar grade silicon workshop, *Memo, 15.05.1999.*
57. TCA termination, Agreement of termination of technical cooperation agreement. parties: Elkem ASA AstroPower Inc, 16.10.2003.
58. C. Dethloff, 2002. *Memo, 25.10.2002.*
59. K. Friestad, 2002. Summary and status of AstroPower 26.06.02, Internal memo, *27.06.2002.*
60. R. Tronstad, 2003. Besøk hos AstroPower, *Memo, 28.01.2003.*
61. C. Dethloff, 2003. AstroPower; Visit, discussion and conclusions, *Memo, 19.08.2003.*
62. C. Zahedi, E. Enebakk, K. Friestad, C. Dethloff, R. Tronstad, K. Peter, and R. Kopecek, 2004. Solar grade silicon from metallurgical route, *Technical Digest of the International PVSEC*, Bangkok, January 26–30.
63. A. G. Forwald and K. Friestad, 2003. Initial solidification tests in induction furnace week 48, *Internal Report F111/2003.*
64. P. Preis, P. Diaz-Perez, K. Peter, R. Tronstad, B. R. Henriksen, E. Enebakk, S. Beringov, and M. Vlasiuk, 2012. High performance solar cells exceeding 17% efficiency based on low cost solar grade silicon, *Proceedings of 7th PVSEC*, Frankfurt, Germany, September 24–28, pp. 1026–1030.
65. Co-operation agreement between BP Solar and Elkem Solar, 17.06.2004.
66. R. Tronstad, 2009. Elkem Solar AS – en ny eksportrettet industri i Agder, *Listerkonferansen*, Lista, Norway, June 12.
67. Meeting Q-Cells-Elkem, Experiments MG-Si 50%/100%, Internal presentation, Amsterdam, 11.12.2006.
68. A. Schei, 2008. Solcellesilisium I Elkem før Elkem Solar, *Presented at Sommermøte I Agder Vitenskapsakademi*, 03.09.08.
69. R. Tronstad, 2012. A sustainable product for the solar silicon market, *Proceedings of the Silicon for the Chemical and Solar Industry XI*, Bergen-Ulvik, Norway, June 25–29, pp. 197–206.
70. R. Birkeland, 2006. Produksjonskampanje solar feedstock, EFS ovn 11 & 12, December 2005, *Internal Report 05.04.2006.*
71. K, Suzuki, K. Sakaguchi, T. Nakagiri, and N. Sano, 1990. Thermodynamics for removal of boron from metallurgical silicon by flux treatment, *Journal of the Japan Institute of Metals*, 54(2), 168–172.
72. A. Schei, 1998. Method for refining of silicon, U.S. Patent 5788945.
73. Norwegian Research Council, Elkem Solar-produksjon av høyrent silisium for bruk i solceller-Fra Sand til Sol, Fact sheet, www.forskningsradet.no.
74. H. Laux, 2003. Final report FSTSII-reactor modelling, *SINTEF Report FSTSII PP01H, 30.12.2003.*
75. K. Friestad, 2002. Improved equilibrium data for a slag treatment of silicon, *Internal Report F128/2002.*

76. J. Heide, 2006. Status og resultater fra slaggproduksjon og slaggbehandling I pyro pilo-tanlegg, *Internal Report, 13.03.2006.*
77. B. Ceccaroli and K. Friestad, 2001. Refining of metallurgical grade silicon, Patent, WO 01/42136.
78. M. Dolmen, 2006. Teknologiverifikasjon for hovedluting iI Solar Prosessen, *Internal Report F139/2006.*
79. R. Tronstad, 2002. New research area—Purification by directional solidification, *Memo, 03.07.2002.*
80. A. G. Forwald and K. Friestad, 2002. Initial solidification tests in induction furnace, *Internal Report F252/2002.*
81. K. Friestad, 2008. Størkning I MMC-S 2008, *Internal Report F28/2008.*
82. R. Gløckner, 2006. Teknologiverifikasjon-etterbehandlingsprosessen, *Internal Report F143/2006.*
83. R. Tronstad, 2004. Scope-Elkem Solar SoG-Si process, *Memo, 09.03.2004.*
84. R. Tronstad and R. Birkeland, 2006. Process description for 5000 MT ES-Si plant, *Memo, 10.01.2006.*
85. Investeringssak: Elkem Solar-5000 MT ES-Si anlegg, *Memo to the Board, 20.10.2006.*
86. Risiko- og sikkerhetsanalyse av industrialisering Elkem Solar, *Rambøl Report, Task 2060096, 28.03.2006.*
87. A. Fredvik, 2009. Elkem Solar AS 5000 MT Industriprosjekt Sluttrapport, August 2009, *31.07.2009.*
88. R. Tronstad, 2009. Elkem Solar—Ground breaking technology for cost leadership, *Presented at NEREC Conference*, Lillestrøm, Norway.
89. P. Galler, A. Raab, S. Freitag, K. Blandhol, and J. Feldmann, 2014. Boron speciation in acid digests of metallurgical grade silicon reveals problem for accurate boron quantification by inductively coupled plasma-optical emission spectroscopy, *J. Anal. At. Spectrom,* 29, 614–622. doi: 10.1039/c3ja50383f, 2014.
90. R. Boisvert, 2008. Timminco Ltd, Bécancour Silicon Inc., *Presented at 6th Solar Silicon Conference*, Munich, Germany, April 1.
91. P. Koven, http://business.financialpost.com/investing/timmenco-files-for-creditor-protection, 03.01.2012.
92. E. Enebakk, A. K. Søiland, J. T. Håkedal, and R. Tronstad, 2009. Dopant specification of compensated silicon for solar cells of equal efficiency and yield as standard solar cells, *Presented at the 3rd International Workshop on Crystalline Silicon Solar Cells, SINTEF/NTNU*, Trondheim, Norway, June 3–5.
93. V. Hoffmann, K. Petter, J. Djordjevic-Reiss, E. Enebakk, J. T. Håkedal, R. Tronstad, T. Vlasenko, I. Buchovskaja, S. Beringov, and M. Bauer, 2008. First results on industrialization of Elkem Solar Silicon at Pillar JSC and Q-Cells, *Proceedings of the 23rd EU PVSEC*, Valencia, Spain, September 1–5, pp. 1117–1120.
94. Technical co-operation agreement between Elkem Solar AS and Calisolar, INC, 30.12.2006.
95. The Kristiansand fairy tale. Norway's Elkem developed a new way to produce solar silicon, *Photon International*, October 2007.
96. K. Friestad, C. Zahedi, M. G. Dolmen, J. Heide, K. Engvoll, T. Buseth, R. Tronstad, C. Dethloff, K. Peter, R. Kopecek, and I. Melnyk, 2004. Solar grade silicon from metallurgical route, *Proceedings of the 19th EU PVSEC*, Paris, France, June 7–11, pp. 568–571.
97. R. Tronstad, 2004. Silisium feedstock for PV industrien, NMS Norsk Metallurgisk Selskap, Presentation at *Metallurgisk Sommermøte*, Trondheim.
98. K. Peter, E. Enebakk, K. Friestad, R. Tronstad, and C. Dethloff, 2005. Investigation of multicrystalline silicon solar cells from solar grade silicon feedstock, *Proceedings of the 20th EU PVSEC*, Barcelona, Spain, June 06–10, pp. 615–618.

99. K. Peter, R. Kopecek, T. Perneau, P. Fath, E. Enebakk, K. Friestad, R. Tronstad, and C. Dethloff, 2005. Analysis of multicrystalline solar cells from solar grade silicon feedstock, *Presented at the Solar Energy Conference*, Orlando, August 6–12.

100. R. Tronstad, E. Enebakk, and J. Vedde, 2006. Silicon production by metallurgical refining process at Elkem Solar AS has shown promising results in multi- and mono crystalline solar cells, *Presented at the International Workshop on Science and Technology of Crystalline Si Solar Cells, CSSC-8*, Sendai, Japan, October 2–3.

101. A. Kränzl, M. Käs, K. Peter, and E. Enebakk, 2006. Future cell concepts for MC solar grade silicon feedstock material, *Proceedings of the 21th EPSEC*, Dresden, Germany, September 4–8, pp. 1038–1041.

102. K. Friestad, E. Enebakk, J. T. Håkedal, and K. Hatlen, 2009. Requirements for silicon used in solar cells, *Presented at the Japan Society of Physics Spring Session*, Tsukuba, Japan, March 31.

103. T. Ulset, 2010. Update on Elkem Solar grade silicon production and application experience, *Photon 8th Solar Silicon Conference*, Stuttgart, Germany, April 27.

104. S. Grandum, A.-K. Søiland, E. Enebakk, and K. Friestad, 2010. Requirements for compensated SoG Si feedstock to be used in high performance solar cells, *Proceedings of CSSC4*, Taipei, Taiwan, October 27–29.

105. T. Ulset, 2013. Elkem Solar Silicon™ staying relevant—Innovation for an oversupplied market, *Presented at the Solar Energy Conference*, Kristiansand, Norway, June 11.

106. H. Aasen, R. Tronstad, and E. Enebakk, 2009. Feedstock specification for multicrystalline silicon solar cells, *Presented at PV-Expo*, Japan, February 25–27.

107. R. Tronstad, A. K. Søiland, and E. Enebakk, 2008. Specification of high quality feedstock for solar cells. *Presented at the Workshop "Arriving at Well-Founded SoG Silicon Feedstock Specifications"*, Amsterdam, Netherlands, November 13–14.

108. C. Dethloff, 2006. Elkem Solar AS. The metallurgical route to solar silicon. *Presented at the 3rd Silicon Conference*, München, Germany, April 3.

109. M. Kaes, G. Hahn, K. Peter, and E. Enebakk, 2006. Over 18% efficient MC-Si solar cells from 100% solar grade silicon feedstock from a metallurgical process route, *4th World PVSEC*, Hawaii, May 7–12.

110. S. Rein et al. 2009. Impact of compensated solar-grade silicon on CZ-silicon wafer and solar cells, *Proceedings of the 24th EU-PVSEC*, Hamburg, Germany, September 21–25, pp. 1140–1147.

111. A. K. Søiland, K. Peter, P. Preis, R. Søndenå, I. Odland, E. Enebakk, and R. Tronstad, 2011. Investigation of CZ-monocrystals, p- and n-type, produced from 50/50 mix of Elkem Solar Silicon® and polysilicon, *Presented at the 5th International Workshop on Crystalline Silicon Solar Cells*, Boston, USA, November 1–3.

112. J. O. Odden, S. Braathen, K. Friestad, M. Tayyib, T. S. Surendra, A. V. Sarma, M. Ramanjaneyulu, R. Nirudi, and T. Ulset, 2013. Superior performance of solar modules based on Elkem Solar Silicon (ESS™) under high solar irradiance conditions, *Presented at Solarcon China*.

113. J. O. Odden, T. C. Lommasson, M. Tayyib, H. Date, R. Tronstad, A. K. Søiland, and T. Ulset, 2013. Results on performance and aging of solar modules based on Elkem Solar Silicon™ (ESS™) from installations at various locations, *Presented at the 2nd Silicon Material Workshop*, Rome, Italy, October 7–8.

114. M. J. de Wild-Scholten, R. Gløckner, J. O. Odden, G. Halvorsen, and R. Tronstad, 2008. LCA comparison of the Elkem Solar metallurgical route and conventional gas routes to solar silicon, *Proceedings of the 23rd EU PVSEC*, Valencia, Spain, September 1–5, pp. 1225–1229.

115. M. J. de Wild-Scholten, R. Gløckner, and R. Tronstad, 2012. Environmental footprint of Elkem Solar Silicon™, Poster at SCSI XI, Bergen-Ulvik, June 25–29.

116. R. Gløckner, J. O. Odden, G. Halvorsen, R. Tronstad, and M. J. de Wild-Scholten, 2008. Environmental life cycle assessment of the Elkem Solar metallurgical route to solar grade silicon with focus on energy consumption and greenhouse gas emissions, *Proceedings of the Silicon for the Chemical and Solar Industry IX*, Oslo, Norway, June 23–26, pp. 235–241.

117. J. O. Odden, G. Halvorsen, H. Rong, and R. Gløckner, 2008. Comparison of the energy consumption in different production processes for solar grade silicon, *Proceedings of the Silicon for the Chemical and Solar Industry IX*, Oslo, Norway, June 23–26, pp. 75–90.

118. A. K. Søiland, G. Halvorsen, K. Friestad, A. H. Amundsen, J. O. Odden, and T. Ulset, 2014. A novel method for improving multicrystalline ingot properties, *Proceedings of the 29th EU PVSEC*, Amsterdam, Netherlands, September 22–26, pp. 536–541.

119. A. K. Soiland, M. G. Dolmen, G. Halvorsen, and R. Tronstad, 2014. Simplified value chain for SoG-Si show promising results at Elkem Solar, *Proceedings of the Silicon for the Chemical and Solar Industry XII*, Trondheim, June 23–26, pp. 241–255.

120. A. K. Søiland et al. 2013. First results from a simplified Elkem Solar route—Input to tolerance limits, *2nd Silicon Materials Workshop*, Rome, Italy, October 7–8.

121. R. Tronstad, 2010. The solar cell industry from a Norwegian perspective, *NEREC Conference*, Lillestrøm, Norway, September 29.

5 From Conventional Polysilicon Siemens Process to Low-Energy Fluidized Bed Processes Using Silane

William C. Breneman and Stein Julsrud

CONTENTS

5.1   TECHNICAL HISTORY OF SILANE PRODUCTION

5.1.1   DEVELOPMENT OF GAS-PHASE PROCESSES FOR PURIFYING SILICON

The challenge to inexpensively purify silicon presents two significantly different routes. The common beginning is the carbothermic reduction of high-quality quartz to produce metallurgical-grade silicon (see Figure 5.1) [1]. The process is energy intensive (14–16 kWh/kg) [2] and the material obtained is typically about 98%–99.5%

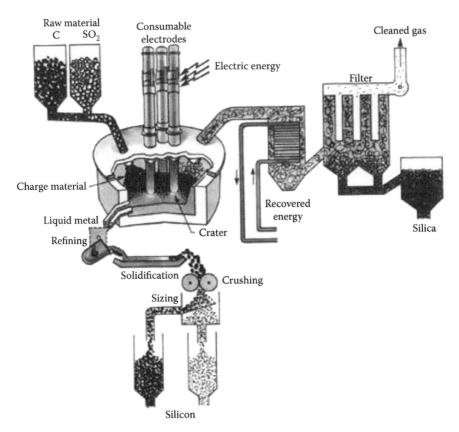

FIGURE 5.1   Carbothermic production of silicon. (Reproduced from A. Schei, J. Tuset, and H. Tveit, 1998. *Production of High Silicon Alloys*, Tapir Forlag, Trondheim, Norway. With permission of the authors and Tapir.)

pure silicon with the major impurities being iron, calcium, and aluminum, with minor amounts of phosphorus, arsenic, and boron.

To refine the silicon to a level suitable for electronic applications, including use in photovoltaic devices, requires at least two orders of magnitude reduction of impurities.

One alternative approach, bulk purification, leads to "UMG-Si" or upgraded metallurgical-grade silicon and is discussed in Chapters 3 and 4. The most vexing problem is to remove, to a very low concentration, and on a consistent basis, the electronically active impurities of boron, arsenic, and phosphorus. These impurities are dissolved in the silicon lattice and are extremely difficult to remove by bulk treatment methods. In contrast, gas-phase purification, the subject of this present chapter, begins with metallurgical-grade silicon (MG-Si) and through a series of chemical reactions converts silicon to volatile silicon-containing fluids. These silicon-bearing fluids are refined by distillation and adsorption techniques to provide ultrapure silicon fluids, either liquid or gas, which are then pyrolyzed in hydrogen to win back an ultrapure silicon product.

Commercial interest in higher-purity silicon developed during the later stages of World War II as a material useful in making high-performance diodes. The invention of the silicon-based transistor in the late 1940s led to a focus on the means to produce, at least on a laboratory scale, silicon with a much higher purity than could be made using the best, at the time, bulk purification means. During this same time period, the silicon industry was developing and the means for preparing volatile silicon materials such as silicon tetrachloride (STC) and trichlorosilane (TCS) were being commercialized. Union Carbide Corp. (UCC), Dow Corning, General Electric, Wacker, DuPont, and other companies were developing a commercial interest in the technology. In the United States, UCC was a firm with broad interests in many areas, including the manufacture of metallurgical silicon and organosilicon materials. It was, therefore, not surprising that UCC would be a major player in responding to the developing need for higher-purity silicon. However, the commercial market for silicon of higher purity than the metallurgical-grade one was just not large enough to attract major commercial commitments. Nevertheless, interest from electronic development laboratories, such as Bell Labs and Fairchild Semiconductor, did create an incentive to at least explore possible routes to higher-purity silicon. Gas-phase purification was seen by visionary scientists as the most plausible means to achieve silicon purity of greater than 99.99%.

5.1.2 HISTORICAL TIME LINE

There are three major time periods in the development of high-purity polysilicon:

1. Mid-1950s through early 1960s. This period saw the initial development of the chemistry and processing to provide the high-purity silicon needed for the developing solid-state electronics industry. The major development was the now classic TCS/Siemens process, which delivered rod-shaped polysilicon from the chemical vapor deposition (CVD) of silicon onto an electrically heated silicon filament.

2. Mid-1970s through mid-1980s. This period was the era of U.S. government sponsorship of the Low Cost Silicon Solar Array Project by NASA/JPL in the United States. Many different methods for purification and deposition were explored, some using refinements of prior technologies and others that blazed new ground. The quest was to demonstrate the production of "solar"-grade polysilicon at a cost of ~$10/kg in 1975 dollars. The motivation was to wean the energy sector of the economy off Middle East oil, which was embargoed in the early 1970s.

3. 2003 through present. International government concern for climate change brought on by the combustion of fossil fuels for energy production led directly to subsidies of photovoltaics and other renewable energy forms and drove rapid expansion of silicon production with a "race to the bottom" for lowest-cost production.

5.1.3   Gas-Phase Purification: Halosilanes

Gas-phase purification of silicon is accomplished by converting MG-Si to a volatile halosilane. The volatile halosilane is then purified by distillation, possibly combined with adsorption, to yield a very-high-purity halosilane. The high-purity halosilane is then reduced to win back the high-purity silicon. Impurities in the form of metal silicides, halides, and hydrides are rejected by distillation or by adsorption. The impurities are treated to be environmentally acceptable. This general chemical purification route can be practiced in various ways for all of the halosilanes.

5.1.3.1   Gas-Phase Purification: Fluorides

Silicon tetrafluoride, obtained as a by-product from the production of phosphate fertilizers, is a low-cost silicon-bearing fluid. The production of SiF_4 removes much of the electronically active impurities. Ethyl Corp. commercialized a process using sodium aluminum hydride to reduce SiF_4 to silane gas [4–6]. The silane is then pyrolyzed in a fluidized bed reactor (FBR) to form the polycrystalline silicon [7]. The major by-product is sodium aluminum fluoride, which was to be sold to the aluminum industry. Ethyl Corp.'s commercial facility in Pasadena, Texas, was sold to MEMC, Inc. in 1988, which was later renamed SunEdison. MEMC also sold silane in the commercial market through distributors.

5.1.3.2   Gas-Phase Purification: Chlorides

Chlorosilanes represent the largest volume and, commercially, the most important volatile silicon-bearing compounds used for the purification of silicon. Chlorides are the "Goldilocks" of the silicon halides. The utility of chlorosilanes lies mainly in their moderate boiling points and chemical stability. The traditional hydrochlorination reaction to form TCS using HCl is carried out in a fluidized bed reactor (FBR) having heat recovery capability. The heat of reaction to produce TCS is 171.14 kcal/mol TCS, which, when generated at 300°C, provides a very available heat source to operate a significant portion of the distillation refining operation. Commercial production of TCS was practiced as early as the 1940s, mainly to provide a feedstock for organofunctional silanes and silicon resins. All of the major silicone (polysiloxane) producers have the capability to produce TCS.

The hydrochlorosilanes can undergo a redistribution reaction whereby TCS can be converted to a mixture of dichlorosilane (DCS) and STC. The DCS can undergo an analogous reaction to form a mixture containing monochlorosilane, TCS, and silane. The composition of the product of the redistribution reaction is governed by the Gibbs free energy of the feed mixture—mostly dependent on the Cl:Si ratio. Various Lewis acid and Lewis base materials will catalyze the redistribution reaction and the kinetics are sufficiently rapid at even modest temperatures such that near equilibrium is achieved in small-sized reactors. This feature, which will be elaborated on later, served as the basis for UCC to enter the Low Cost Silicon Solar Array Project "contest" in the mid-1970s. But the traditional production process using TCS has been, and is still, the dominant technology since the mid-1950s. TCS of sufficient purity to allow the production of "electronic" or "solar" grades of silicon by hydrogen pyrolysis of the TCS is the fundamental basis of the high-purity silicon business.

5.1.3.3 Gas-Phase Purification: Bromosilanes and Iodosilanes

The bromosilanes and iodosilanes have been studied and process routes proposed to purify MG-Si to high-purity silicon in closed cycle processes [8]. While the bromo- or iodosilane processes have some claimed advantage relative to process simplicity and efficiency in the removal of impurities, they have not found significant commercial favor due to the environmental problems of disposing of bromine or iodine salts of the impurities in the MG-Si. Bromosilanes and iodosilanes are also very corrosive to iron-based alloys unlike the chlorosilanes, which can be handled quite well for processing chlorosilanes at temperatures up to about 400°C.

5.1.4 POLYSILICON: TCS SIEMENS ROUTE

High-purity silicon is produced today predominantly by the "TCS-Siemens" route. MG-Si is crushed to a size suitable for use in an FBR. At a modest pressure of about 3 bar, anhydrous hydrogen chloride fluidizes the silicon mass and at a temperature of about 300°C reacts exothermically with silicon to form a mixture of TCS, STC, and hydrogen:

$$Si + 3.2HCl \rightleftharpoons 0.8HSiCl_3 + 0.2SiCl_4 + 1.2H_2 \qquad (5.1)$$

These ratios of TCS and STC are close to the actual reaction products but can vary with reactor temperature and the presence of impurities associated with the silicon raw material or additives purposely employed to shift the product mixture. The reaction is not allowed to proceed to thermodynamic equilibrium, but is quenched by cooling.

TCS was initially produced as a precursor for organofunctional silanes. Organofunctional silanes find use as coupling agents to promote a chemical bond between glass fibers and polymeric matrixes to add strength and water resistance to fiberglass composites.

By fractional distillation, TCS could be refined to high purity. The most troublesome impurities to remove are boron trichloride, phosphorus trichloride, and certain

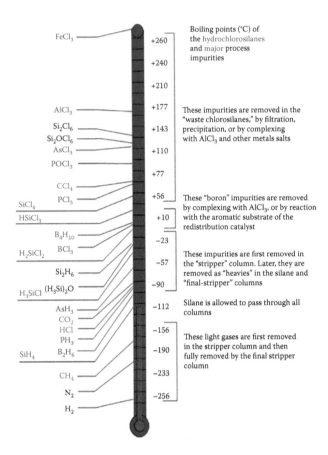

FIGURE 5.2 Boiling points of relevant chemical species present in the production of silane.

hydrocarbons that have boiling points close to TCS (see Figure 5.2). When coupled to a facility that also produces organofunctional silanes based on TCS, a means is provided to reject somewhat impure TCS as an "internal" grade suitable for the organofunctional products. The importance of this "blow-down" stream cannot be overemphasized. In order to remove boron impurities, chemisorption onto silica gel is generally used while some producers used a proprietary method involving moisture in controlled amounts to form low-volatility borates from the BCl_3. PCl_3 and higher-boiling phosphorus chlorides and hydrocarbons are generally removed to a satisfactory level by distillation—especially if a purge of impure "internal" grade of TCS can be utilized.

With high-purity TCS available, solid polycrystalline silicon was produced by pyrolysis in hydrogen. Early processes used a simple quartz tube in a furnace wherein the silicon collected on the inside of the tube. Once the tube became filled, the whole tube was sacrificed, the quartz was etched away with acid, and the shiny silicon chunks were packaged for sale. In the Siemens process, the reactor consists of a series of electrodes that hold slim seed rods of silicon that are directly heated by an electric current. Siemens, having primarily an electric controls focus, developed several methods to control the heating of the seed rods. Pure silicon seed rods have a high resistivity when cold, but the resistance drops sharply as the silicon gets hot. Special electric power controls or auxiliary heating is needed to control the power applied to the seed rods to prevent them from melting during the initial heat up. Initially, the seed rods were contained within a quartz bell jar and the deposition was continued until the rods grew to the limits of the electrical power supply. The beauty of the Siemens process is that the product silicon is not touched or contaminated except during the "harvesting" exercise at the end of the run, and any minor surface contamination could be washed off by an acid etch and rinse. The TCS/Siemens process thus became the industry standard for the production of high-purity poly-crystalline silicon from about 1960 until the present day.

5.2 SILANE DEVELOPMENT

5.2.1 LITHIUM HYDRIDE REDUCTION

Silane has been made by several synthesis methods. In the Litz-Ring [9] process, lithium hydride is used to reduce STC in a lithium chloride/potassium chloride molten salt mixture at 400°C:

$$4LiH + SiCl_4 \rightarrow 4LiCl + SiH_4 \tag{5.2}$$

This reaction is suitable for small-volume production, and was in commercial practice by the Linde Division of UCC in 1975. However, for larger-scale production, conversion of the LiCl by-product using electrolysis followed by recycle of the chlorine to react with MG-Si would be needed. These later steps were not commercialized.

5.2.2 MAGNESIUM SILICIDE ROUTE

Silane has also been prepared in commercial quantities by the reaction of magnesium silicide and ammonium bromide. First, magnesium silicide is formed by the exothermic reaction of magnesium and silica [10]:

$$4Mg + SiO_2 \rightarrow 2MgO + Mg_2Si \tag{5.3}$$

The magnesium silicide is then reacted in a liquid slurry of ammonium bromide:

$$Mg_2Si + 4NH_4Br \rightarrow 2MgBr_2 + 4NH_3 + SiH_4 \tag{5.4}$$

The magnesium silicide route was used by Komatsu Electronic Metals (KEM) beginning 1966 in their commercial production of polycrystalline silicon. One important feature of the process was that it produced silane having extremely low levels of boron as any diborane (or other boron compounds) would react readily with the liquid ammonia. KEM used their own unique "Siemens"-style reactors [11], which were specially modified to minimize the formation of silicon powder.

It was however not until the mid-1970s that a low-cost, high-volume process for producing silane was conceived and developed by UCC under a contract with the Jet Propulsion Laboratory and the Department of Energy [12].

The development of a process to produce high-purity silane as a precursor to polysilicon was not linear but rather circular. In the 1950s, UCC was using TCS to produce organo-modified siloxane resins, which made a reasonably high-purity TCS available. UCC was also interested in producing DCS from which diethylsilicones could be made as a high-performance silicone brake fluid that had the desirable feature of being "paintable" unlike dimethylsilicones. Dimethylsilicones were the most common silicones thanks to the discovery by Rochow [13] that allowed dimethyldichlorosilane to be produced in good yield directly from methyl chloride and copper-catalyzed MG-Si. But DCS was not able to be made in significant yield by the direct reaction of ethyl chloride and MG-Si, even with copper catalyst. An amine-based redistribution process was invented [14] to convert TCS to a mixture containing up to 15% of DCS. Reaction of DCS with ethylene yielded diethyldichlorosilane, which by hydrolysis resulted in an ethylsiloxane fluid. Such "ethylsilicones," unlike the more common methylsilicones, can be easily painted over. But the product was never able to be made reliably in a manner to satisfy the automotive industry's criteria for paintability and the technology development path stalled.

In the early 1970s, DCS was of interest as a silicon-bearing gas, which could be used to improve the growth rate and yield in a conventional Siemens-style reactor [15]. A project was initiated at UCC's Sistersville research center to produce a small amount of DCS using the Bailey redistribution process. At the same time another project was ongoing to facilitate the production of cyanopropyltrichlorosilane by the coupling reaction of acrylonitrile and TCS. It was found that an amine functional ion exchange resin would catalyze the reaction, but unfortunately also resulted in only about 80% yield due to the formation of STC and a higher-boiling adduct of acrylonitrile to DCS [16]. The chosen resin catalyst was active for the redistribution of hydrochlorosilanes and the solid catalyst made a continuous process possible. Quickly, a small contact bed of the catalyst was put into operation to treat a tank wagon full of TCS and produce a tank wagon of a mixture of DCS, TCS, and STC. The DCS was recovered using an available distillation unit. This resulted in the first commercial production of electronic-grade DCS. A laboratory setup was simultaneously used to demonstrate the first truly "reactive distillation" production of silane by using an amine functional weak base macro-reticular styrene-divinyl benzene ion exchange resin catalyst as the vapor–liquid contact surface in a distillation column. TCS was refluxed through the packed column and silane was vented from the refrigerated condenser [17].

5.2.3 JPL Program for Silane

In 1975, as a result of a national energy policy, a development program was established to foster the technology for producing low-cost silicon solar cells, which would be suitable for large-scale power generation [18]. The program was managed by the Jet Propulsion Laboratory as they were experienced in the technology of photovoltaics. UCC was one of 11 participants focused on developing a process for the high-volume, low-cost production of solar-grade silicon. UCC proposed a process that was based on the work of Literal and Bakay. The argument put forth to the JPL contract managers was that while the proposal did not specifically address the production of solar-grade silicon, the proposed process would yield silane of exceptionally high purity from a process that would be readily scalable with little risk of "failure." According to Don Bailey, the UCC manager, it would be "a lead pipe cinch." And given the low cost and high purity expected of the silane, there were several known pathways open to convert silane to silicon. The process would utilize the solid amine functional resin to catalyze the redistribution of hydrochlorosilanes (TCS and DCS) and standard fractional distillation to separate the hydrochlorosilanes.

The lower-boiling chlorosilanes (see Figure 5.2) would be passed to the down-stream distillation columns and an additional redistribution reactor(s) until silane was recovered, while the higher-boiling fractions would be passed to the beginning of the distillation operation where STC would be rejected. The co-product STC would be sold (utilized) as a commercial raw material into an established market.

The purity of the silane was expected to be very high since the boiling points of the known or suspected impurities was dramatically different from silane ($-112°C$) (see again Figure 5.2). The major process design decisions focused on the operating pressure in the various reaction and separation steps to minimize the expense of low-temperature refrigeration and to maximize the benefits of the resin catalyst.

Shortly after the JPL development contract was put into execution, it became apparent that the volume of STC, which would be generated for large-scale silane (silicon) production, would seriously upset the STC commercial market. For every kilogram of silane produced, 16 kilograms of STC are also produced. The stoichiometry is straightforward:

$$2HSiCl_3 \rightleftharpoons H_2SiCl_2 + SiCl_4 \tag{5.5}$$

$$2H_2SiCl_2 \rightleftharpoons H_3SiCl + HSiCl_3 \tag{5.6}$$

$$2H_3SiCl \rightleftharpoons H_2SiCl_2 + SiH_4 \tag{5.7}$$

Overall:

$$4HSiCl_3 \rightarrow SiH_4 + 3SiCl_4 \tag{5.8}$$

A practical chemistry and process was needed that would convert STC back to TCS (and hopefully) consume MG-Si:

$$3SiCl_4 + 2H_2 + Si \rightleftharpoons 4HSiCl_3 \tag{5.9}$$

Within the archives of UCC's technology, there was a patent that described this reaction in quite a detailed manner [19]. Interestingly, no further mention was made in the UCC literature about this reaction as the commercial focus was on the production of TCS by reaction of HCl with MG-Si.

5.2.3.1 Redistribution Chemistry

The redistribution reactions of hydrochlorosilanes, which form one of the two major chemistries of the silane process, are based on the chemical equilibrium of the H–Si–Cl system. The composition of the various hydrochlorosilanes is a function of the chlorine/silicon ratio and the process temperature (see Figure 5.3).

At a typical process temperature of 70°C, the equilibrium compositions are as shown in Figure 5.3.

The reaction kinetics for the redistribution reaction using a macro-porous substrate for the weak base catalyst (DOWEX® MWA-1, Amberlyst® A-21, or similar) was measured for TCS and DCS feed materials. For DCS vapor, the reaction reached substantial equilibrium in less than 10 s contact time [12]. The resin catalysts selected for the JPL project work were commercially available. The properties were well characterized by the vendors. The typical commercial material however is manufactured in an aqueous suspension and thus the resin must be made anhydrous prior to use in the chlorosilane process. The initial approach was to solvent exchange using first alcohol to remove the water, then a light hydrocarbon (toluene, octane, etc.) to remove the alcohol, and finally STC to remove the hydrocarbon. All of this resulted in large amounts of hazardous waste. An alternative was to thermally dry the resin with warm air at 80–100°C. This was very successful and a stable, highly active, and long-lasting catalyst resulted. The actual longevity of the resin catalyst has not been determined. While the initial activity was reduced, after a short period of time, the reactivity stabilized for such a long time that a real "lifetime" has never been determined. Commercial experience indicates that factors other than chemical reactivity will result in a need to change the catalyst.

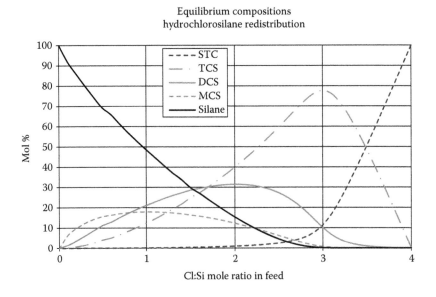

FIGURE 5.3 Equilibrium distribution of hydrochlorosilane in the product mixture as a function of feed composition.

5.2.3.2 Hydrogenation Chemistry

The term "hydrogenation" was first applied to the chemistry of the reaction

$$2H_2 + 3SiCl_4 + Si \rightleftharpoons 4SiHCl_3 \tag{5.9}$$

by UCC, but a perhaps more correct term would be hydrochlorination (see Section 5.9). Regardless, for this chapter, the term "hydrogenation" will continue to be used to differentiate it from the reaction of silicon with HCl:

$$Si + 3.2HCl \rightleftharpoons 0.8HSiCl_3 + 0.2SiCl_4 + 1.2H_2 \tag{5.1}$$

The yield in the "hydrogenation" reaction is limited by chemical equilibrium. But the kinetics are governed by the surface chemistry of the MG-Si as well as the reaction temperature and system pressure. The reaction is favored by higher pressure since there are fewer moles of TCS in the product than moles of gas in the feed. Higher pressure also has an impact on the physical size of the reactor. In laboratory studies, the system pressure was varied from near atmospheric up to about 37 atmospheres [20,21]. Details of the reaction kinetics are shown as an example in Figure 5.4.

5.2.3.3 Effect of Catalyst

The kinetics of the "hydrogenation" of MG-Si with H_2 and STC are strongly affected by the surface chemistry and intergranular features of the silicon charge. Copper has

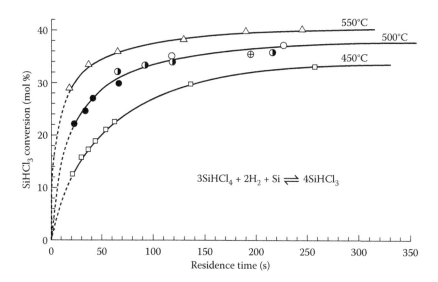

FIGURE 5.4 TCS hydrogenation kinetics. (Adapted from J. Mui and D. Seyferth, 1981. Document DOE/JPL 955382-79/8(DE81025543).)

been known since the work of Rochow in 1940 [13] to have a profound effect on the reactivity of silicon. The early work by Wagner of UCC also highlighted the role of copper in promoting the reaction of STC and H_2 with silicon [19]. More recently, Mui reaffirmed the importance of copper in this reaction [20]. A mechanistic study of the role of copper in the reaction was presented by Röwer [23], which indicated that while iron, nickel, and cobalt provide some catalytic activity with silicon, the best performance from a theoretical view was copper. The overall reaction is believed to include electron transfer steps from the metal silicide catalyst to adsorb STC molecules. A silylene species ($SiCl_2$), which is formed from $SiCl_4$ and the catalyst surface is proposed. A hydrogen molecule injects electrons into the solid, whereby hydrogen chloride is generated on the surface. TCS results from the oxidative addition of HCl to chemisorbed $SiCl_2$ [24]. The mechanism is operable at temperatures greater than about 400°C.

Using preliminary reaction kinetics and equilibrium data, a small "mini-plant" was designed, fabricated, and operated under the JPL contract to demonstrate the feasibility of producing silane from MG-Si at a quality suitable for solar-grade silicon. The mini-plant is shown in Figures 5.5 and 5.6.

FIGURE 5.5 Mini-plant for silane production.

FIGURE 5.6   Mini-plant: Silane loading station.

The demonstration for silane production was successful. The product was tested by KEM by growing silicon from the silane in their small CVD system and analyzing for its electronic properties. It was found suitable for growing solar-grade silicon material.

A manufacturing cost estimate (Table 5.1) based on the mini-plant's operation showed the potential for silane to be produced at about $4.35/kg (1977 basis) on a variable (cash) cost basis. The manufacturing costs were estimated considering first

TABLE 5.1
Preliminary Silane Cost Estimate for the SiCl$_4$ Hydrogenation Route

Quantity	Item	Unit Cost ($)	Cost ($)/Pound
1.09 Lb/Lb SiH$_4$	Silicon	0.7235	0.7886
22.4 (Cuft)	Hydrogen	0.01	0.224
1.12	Hydrogen chloride	0.07	0.0784
0.02	Copper	1.21	0.0242
47.04	Steam (redistribution)	0.00407	0.19145
0.307 (kWh)	Refrigeration	0.018	0.0055
22.98	Steam recycle cost	0.00407	0.0935
Total variable costs ($)			*1.4057*
0.00714	Two operating positions at $40/h	80	0.571
Total period cost			*0.571*
Total cost per pound of product ($)			*1.977*

Source: Adapted from W. Breneman, 1977. *Low Cost Silicon Solar Array Project.* Project 486S80 File No. SVP-77-1, Union Carbide, Sistersville, WV.

the hydrogenation of SiCl$_4$ to HSiCl$_3$ at 550°C using Cu as the catalyst (20% conversion), followed by redistribution to SiH$_4$ at a production rate of 140 pounds/h of SiH$_4$.

5.2.4   Process Design for a 100 Ton/Year Demonstration Plant

The next phase of the JPL research contract called for construction of an experimental process systems development unit or EPSDU. During this phase, the physical properties of the hydrochlorosilanes were updated, the materials of construction for the various unit processes were defined, and a complete process package was developed. Fortunately, the properties of the hydrochlorosilanes did not show significant deviation from ideality and that allowed standard chemical engineering process simulation software to be utilized to define the stream and energy flows.

The EPSDU (Figures 5.7 and 5.8) constructed at a UCC location in Washougal, Washington, was designed to be a small-scale (100 ton/year) version of what was hoped to be a full 1000 metric ton (MT)/year commercial plant.

An important part of the process design was the selection of the process operating conditions. Operating pressures for the various unit operations were carefully selected to minimize the cost of necessary refrigeration. The temperature range for the redistribution reactors needed to consider the stability of the catalyst. The operating pressure and temperature of the hydrogenation reactor was selected in consideration of the materials of construction and physical size of the vessel. The EPSDU operated from January 1983 until February 1986. The results of its operation were documented in a final report to JPL. The more significant findings were as follows:

1. Mechanically sealed pumps in chlorosilane service experience rapid seal failure due to poor lubricity of the chlorosilanes and the negative effects of chlorosilane contamination of the lubricating oil.

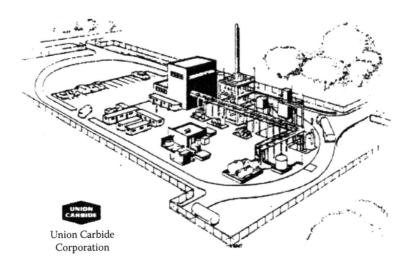

Union Carbide
Corporation

FIGURE 5.7   Experimental process system development unit.

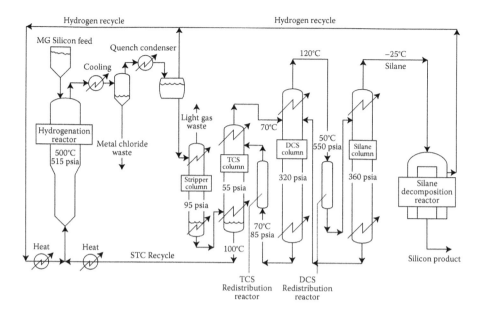

FIGURE 5.8 Process flowchart for the EPSDU silane plant.

2. Improper gas distribution in the fluidized bed hydrogenation reactor can very seriously reduce reactor effectiveness.
3. Vaporizers and boilers processing chlorosilanes can experience severe fouling if the system is contaminated with pump seal oil.
4. Regardless of the problems of contamination of the chlorosilane streams, silane of very high quality would be produced.

5.3 PRESENT-DAY SILANE PROCESS

Today's collection of silane production facilities operated by REC Silicon includes the original Moses Lake plant with a capacity of 2600 MT/year of electronic- or solar-quality silane, a 5600 MT/year facility in Butte, Montana, producing "electronic"-quality silane, and two 9600 MT/year production units at Moses Lake (see Figure 5.9), each capable of producing electronic-grade silane.

The basic process is very similar to the original EPSDU. Alterations to the process were mainly focused on site-specific waste treatment units, revised means for condensing the chlorosilanes from the hydrogen atmosphere in the hydrogenation unit in a manner to minimize fouling from aluminum chloride complexes, and improved energy capture from hydrogenation reactor exit gases. To provide higher-quality silane for the commercial market, an "after stripper" column was added to the distillation train to remove any low-boiling "permanent" gases from the silane that may have entered the process, such as from maintenance service activities or other non-routine events. Boron and phosphorus levels in the final product are well under the published specification of 10 parts per trillion boron and 20 parts per trillion phosphorus.

FIGURE 5.9   REC silicon production plant at Moses Lake, Washington. (REC Silicon, Inc., 2014. *REC Annual Report,* REC Silicon, Moses Lake, WA.)

5.3.1   CORPORATE DEVELOPMENT

UCC's technology and commercial agreement with KEM came to a close in 1992 when KEM bought out UCC's complete interest in the polysilicon business and the facility in Moses Lake. KEM expanded the polysilicon operation by building a new plant in Butte, Montana in 1999, identifying it as Advanced Silicon Materials, Inc. (ASiMI). KEM in turn exited the polysilicon business in a series of financial moves whereby Renewable Energy Corp. (REC) took over ownership of both the Moses Lake and Butte, Montana, plants in 2005.

REC was a vertically integrated business that originally included polysilicon production through to production of complete photovoltaic modules and systems. In 2010, REC divested the cell/module business from the polysilicon production. The silane/polysilicon production at Moses Lake and Butte was retained under the name of REC Silicon and the wafer-to-module production in Singapore was retained under the name REC Solar (see Chapter 4). The worldscape of silane production now has several recent entries: OCI Materials and SMP, a joint venture in Korea of MEMC/SunEdison and Samsung among others [27]. They all use the basic silane process described earlier [28]. Complete process design packages are available from, among others, GT Advanced Technologies (GTAT) for silane production [29].

5.3.2   SILANE PRODUCTION COSTS

The "cash cost" for producing silane is not publically available. But reasonable assumptions can be made if the concept of a closed cycle process is used. The major raw materials are MG-Si, hydrogen (net of any recycle from the polysilicon production system), and either STC or anhydrous HCl. MG-Si contains up to 2% impurities, some of which are converted to their chlorides, while others are manifest as silicides or other unreactive materials. The fraction of MG-Si not converted to useful silanes must be treated for proper disposal. STC or HCl are used to replace the losses of chloride and alkali; lime is preferably used to neutralize the acidic chlorides. The costs that pass to the silane are a reflection on the efficiency of the process design and plant operations to minimize the quantity of STC/HCl makeup. In addition, a neutralization agent, such as caustic or lime, is needed to neutralize any HCl generated in the waste treatment area, process vent scrubbers, etc. The prices of these commodities are publicly available. Energy utilization is primarily in the form of process heat, with a minor amount devoted to gas compression, fan motor power, and fluid pumping. The heat energy is largely interchangeable between direct electric heat and conventional combustion. Depending upon the location, direct electric heating can have a much lower life cycle cost than fossil fuel, especially for high-temperature process requirements. There is also opportunity for interprocess energy recuperation. As with any industrial process, there is a trade-off between the capital investment of the energy recovery equipment, the energy savings, and the operability of the process. The silane process is very amenable to computer-based modeling. As such, the energy flows can be determined and the process optimized.

5.3.3 SILANE PURITY

The purity of commercial silane is excellent and superior to the requirements for "solar-grade" polysilicon. Typical silane specification and assay are listed in Table 5.2 [30]. The electronic qualities of silane are determined by converting the silane to silicon in a quality-controlled CVD reactor and then measuring the electrical properties and compositions thereon of the produced silicon (see Table 5.3).

TABLE 5.2
Bulk Silane Specifications

Impurity Component	Maximum Concentration (ppmv)	Test Method
He	4	Gas chromatography
Ar	0.06	Gas chromatography
Disiloxane	0.1	Gas chromatography
C_2H_6, C_3H_8, C_4H_{10}	0.15	Gas chromatography
CO	0.2	Gas chromatography
CO_2	0.1	Gas chromatography
H_2	30	Gas chromatography
Total chlorosilanes	0.1	Ion chromatography
Si_2H_6	1	

Note: ppmv = parts per million in volume.

TABLE 5.3

Analysis of Polysilicon Deposited from Silane in Parts per Trillion Atomic (ppta)

Element	Concentration	Method of Detection
C	<0.1 ppma	FTIR
P	<20 ppta	FTIR/FTPL
B	<20 ppta	FTIR/FTPL
Al	<5 ppta	FTIR/FTPL
As	<5 ppta	FTIR/FTPL
Ga	<5 ppta	FTIR/FTPL
Sb	<5 ppta	FTIR/FTPL
In	<5 ppta	FTIR/FTPL
Total metals	<0.4 ppta	NAA

Note: Resistivity >10,000 $\Omega \cdot$ cm.

Fourier transform infrared spectroscopy (FTIR) and Fourier transform photoluminescence spectroscopy (FTPL) are two highly sensitive techniques used to determine the low levels of boron, phosphorus, aluminum, gallium, arsenic, indium, and antimony in silicon, while neutron activation analysis (NAA) is used for other metals. Gas chromatography is used to quantify the volatile impurities.

5.3.4 SILANE PROCESS SAFETY

Silane is dangerous and the release of silane into the air can lead to an explosion without warning and can result in devastating loss of life and property. However, the flammability characteristics of silane are pretty well known and it will not explode without warning [31]. In over 30 years of producing silane in large volumes, REC Silicon has not had a major incident involving silane. Silane has a flammability envelope in air from about 1.5% up to over 90%, very similar to hydrogen. When silane burns, it produces white silica and water vapor, except when there is a shortage of oxygen, when the products of combustion are amorphous silicon and hydrogen. Silane released from a high-pressure source may not ignite immediately, except if the jet is disturbed by impingement on a solid baffle or the jet picks up the relatively warm solid's (compared to the jet's temperature of −112°C) temperature [32]. Extensive tests have shown that large-volume releases are actually less hazardous than small leaks within an enclosed space (gas cabinets are a prime example) where an ignition delay can result in an accumulation of a metastable mixture of silane in air. This metastable mixture can react violently once the concentration exceeds a threshold value of about 4%. The ensuing rapid pressure rise can destroy the enclosure.

Guidelines have been developed over the years to facilitate safe handling of silane in bulk containers of up to 12 tons. Those containers have been subjected to the rigors of transoceanic travel without serious effect and loss of life or property in over 30 years.

In the production plants, certain guidelines for silane process safety are rigidly followed:

- Piping is of a "diffusion-resistant" character emphasizing metal-to-metal joints.
- Vessels are designed to ASME Sec VIII standards with pressure ratings conservatively assigned.
- Silane is stored in bulk for in-plant use as a refrigerated liquid under pressure.
- Silane is moved within the facility as a gas, not a liquid.
- A rigorous mechanical integrity program is in place.
- A proven method is used to remove traces of silane from process equipment prior to opening the equipment for service—this includes small valves and instrumentation.
- Preparing equipment for operation follows the same rigor as the preparation for service, except in reverse.
- Operation staff is well trained in the approved procedures and retraining is frequent.

5.3.5 SILANE STORAGE

To provide greater assurance of silane supply for the silicon production units, silane can be inventoried in either high-pressure gas storage vessels or as refrigerated liquid under more modest pressure. While there are advantages to both, the economics usually favor the liquid storage method. In either case, reheaters are needed on the silane storage vessel discharge lines to raise the gas temperature. With the silane distribution system operating at pressures less than 20 atm, the temperature of the gas would otherwise drop to silane's boiling point at the low delivery system's pressure (Figure 5.10). Storing silane as a refrigerated liquid under pressure requires a continuous electrical power supply for refrigeration. Silane's critical temperature is −4°C. Loss of refrigeration means that silane must be vented from storage at a rate where the heat of vaporization is balanced by the heat gain through the storage vessel's insulation. Only in the event of a total unplanned plant shutdown of significant duration would silane need to be vented to maintain safe operating pressure in the storage tanks.

5.3.6 OPERATIONAL RELIABILITY

Operational reliability is a significant issue in the commercial-scale plants handling chlorosilanes at elevated pressures and temperatures. An intensive program of mechanical inspection must be in place. Constant vigilance is needed to identify and correct minor fugitive emissions. Atmospheric HCl generates when a chlorosilane leak takes a toll on ferrous materials, even if they have a protective coating (paint). This is especially critical if a leak occurs under insulation where it survives undetected until serious corrosion has occurred. In critical service, normal austenitic

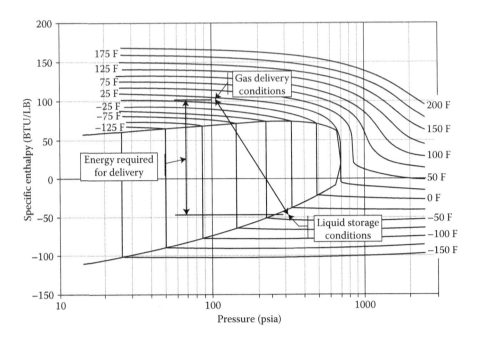

FIGURE 5.10 Pressure–enthalpy–temperature chart for silane. (Adapted from R. B. Richardson, 1984. *Thermodynamic Package SGS-69*, SGS-69, Union Carbide Corp., Washougal, WA; REC Silicon, Inc., 2014. *REC Annual Report*, REC Silicon, Moses Lake, WA.)

steels are replaced with nickel alloys. All of these factors contribute to a higher than average cost for plant construction and maintenance overhead.

The silane production process consists of two major process units—the conversion of MG-Si to liquid chlorosilanes and the conversion of that feedstock to silane. Without one, the other is quickly out of raw material. Beginning with the first plant at Moses Lake, REC Silicon's philosophy has been to install silane production facilities in pairs. An outage at one unit will not completely shut down polysilicon production. Polysilicon production relies upon a rather steady and predictable flow of silane and simultaneous consumption of the co-product hydrogen. Today, a single silane production plant would have a capacity of about 12500 MT/year of silane [34]. The REC Silicon/Shaanxi Non-Ferrous Tian Hong New Energy Co joint venture plant being built in Yulin, China, continues this present scale of operation with two 12500 MT/year silane production units that will together produce 19000 MT/year of granular silicon, 1000 MT/year of Siemens-grade polysilicon, and 500 MT/year of merchant silane [35].

5.3.7 SILANE MERCHANT MARKET

Silane is sold internationally in bulk high-pressure tube modules. The largest have a capacity of 6 MT of silane in a 40 ft rated module. The global merchant market for silane in 2014 was about 4000 MT.

5.4 TECHNICAL HISTORY OF SILANE-BASED POLYSILICON PRODUCTION

5.4.1 SILANE-BASED METHODS OF POLYSILICON PRODUCTION

The deposition of silicon from silane is essentially an irreversible, thermally initiated, decomposition reaction, which yields only silicon and hydrogen:

$$SiH_4 \rightarrow Si + 2H_2 \tag{5.10}$$

The silane decomposition reaction self-initiates at a temperature in excess of 300°C. In reality, the reaction is much more complex [36]. Bulk gas decomposition occurs at a relatively low temperature and produces fine particles of amorphous silicon. Decomposition near a hot surface (750°C) will form a dense coating of polycrystalline silicon. Rapid quenching of a decomposing silane stream leads to the formation of polysilanes. The designer must choose the thermal and time environment depending upon the desired outcome and arrange the process environment accordingly.

5.4.2 ROD-LIKE POLYSILICON

Pyrolysis of silane to form dense polycrystalline silicon is done in a "modified" Siemens-type reactor [11]. Commercial-scale reactors were developed by KEM in the late 1960s. The rods are heated by electric current in the same fashion of the traditional Siemens reactors. The rod temperature is generally much lower (800°C) than is the case with the chlorosilanes (1000°C). A silane-based polysilicon reactor is characterized by having water-cooled "pockets," which surround each seed rod. Like TCS-based reactors, strong gas convection flows move the gas atmosphere upward along the hot silicon rods and down along the cooled walls. By keeping the bulk gas temperature cool, formation of amorphous powder is minimized. However, with the greatly increased gas cooling, and high surface area of cool walls surrounding each heated polysilicon rod, radiation losses are quite high. The power consumption is increased such that the overall power consumption per kg of silicon produced is substantially equivalent to that of TCS-based reactors operating to produce the same product form. Unlike a TCS-based reactor, pure silane is the feed gas. No hydrogen is fed during the deposition period as the hydrogen that is formed by the decomposition is sufficient to provide the convection currents for cooling the bulk gas phase. Increasing the flow rate of silane at a constant reactor rod temperature results in an increase in the rate of polysilicon deposit. However, this is balanced by changes in the morphology of the polysilicon rod and the co-generation of fine powder silicon. Some of the powder is attracted to the cold walls of the reactor by thermophoresis. Balancing the cooling rate, silane feed rate, and rod temperature is an art developed over the years to optimize the production of the desired product form and under the most economical conditions.

5.4.3 NON-ROD REACTOR DEVELOPMENT

The development of silane-based silicon production at UCC began at the Carbon Product's Division Research Laboratory in Parma, Ohio, in the early 1970s.

The emphasis was always on producing an "electronic"-grade polycrystalline silicon product. Initially, there were two development paths: a free space reactor (FSR) and an FBR.

The FSR was the more technically developed by the mid-1970s and appeared to pose a lower hurdle to become commercialized. The FSR consisted of an empty vertical cylindrical reactor with a quartz liner. It was heated externally and silane (perhaps mixed with hydrogen as a diluent) was fed through the center top of the reactor. A brown amorphous silicon powder was collected in a gas/solid disengaging space below the reactor. In the laboratory version, the silicon powder was simply harvested from the powder hopper and, under a controlled atmosphere, fed to a heated quartz crucible where it was melted and solidified into a shape suitable for analysis or other purposes [37]. In 1977, the FSR and the FBR development programs were added into the JPL research program on silane. Interest continued in the FSR, and the technology was developed at a pilot scale to the point where the amorphous silicon powder was demonstrated to be collected. The powder was transported 60 miles to another location where a facility was installed to melt the powder in a modified Czochralski crystal puller furnace. The modification consisted of mounting the furnace atop a "shotting tower." Molten silicon was flowed from the bottom of the crucible and allowed to drip into the tower to a collection vessel below. The silicon formed a nominally 3 mm round "shot" [38]. While the essential features of the FSR/powder melting/shotting process were demonstrated, the project was shelved in favor of the FBR, which appeared to offer a path to a greater production capability.

5.4.4 FLUIDIZED BED DEVELOPMENT

A fluidized bed method for depositing silicon from a high-purity silicon-bearing gas is very attractive from an economic standpoint. Fluidized beds can have a very low cost relative to their capacity and product uniformity should be quite high. They also should present a low energy footprint. Texas Instruments attempted to produce polycrystalline silicon granules from TCS using a fluid bed technique in the 1970s [39]. Several other firms have attempted this process. To date, only Ethyl Corp (now SunEdison) and REC have had significant commercial success. One of the big technical challenges is to preserve the purity of the high-purity silicon-bearing feed gas. The advantage of using silane as the source gas substantially eliminates two major concerns. First, silane has no chloride and thus the potential for corrosion is greatly reduced. Second, there is no chemical equilibrium to contend with; the reaction is unidirectional and the products of the decomposition are only silicon and hydrogen. UCC's and JPL's FBR programs recognized these advantages and explored the challenges posed by the other feature of silane thermal decomposition: that the pathway to a dense crystalline form of silicon is complicated with many reaction pathways that lead to silicon. Deposits took forms from amorphous, nanoscale dusts to multigrained crystalline forms, which agglomerated together into an intractable mass. When coupled with a desire to maintain the high purity of the silane raw material, the challenge was even greater. UCC's focus was always on the manufacture of the highest-purity material. That philosophy guided the research effort and largely excluded investigation into materials of construction other than quartz.

The continued focus on electronic quality significantly hindered the development of a practical FBR.

The fluidized bed research program under UCC was temporarily halted when JPL funding terminated in the mid-1980s. Meanwhile, Ethyl Corp. was successful in commercializing a silane-based FBR at their plant in Pasadena, Texas [40]. Ethyl was producing silane via a different synthesis route, but it was their fluid bed silicon product that was the center of attention. Ethyl's silane/FBR unit was sold to Albemarle and then taken up by MEMC in the early 1990s. Their FBR product has been on the commercial market since the late 1980s. But the purity was not too good and as a result its acceptance in the market was hindered. In 1992, UCC sold their silicon business to KEM who renamed the silane–silicon business ASiMI. ASiMI rekindled the fluid bed program in 1996 with funding by the new owners. The major investment was the construction of a tall pilot plant building to house an all-quartz reactor, in keeping with the original objective to produce an electronic-quality silicon product. But the continued focus on quartz was also a millstone on the project as the reactor parts were prone to breakage. In 2002, REC bought 50% of Moses Lake facility from ASiMI, changed the focus of the FBR to a "solar" grade of silicon and altered the FBR to one made of metal, and named it FBR-A when it was commercialized in 2009. The change to a metal reactor allowed more rapid development and shortly the major classic issues of excessive powder formation, wall deposits, and agglomeration within the reactor were reduced to manageable levels. Part of the work was aided through the use of computer-aided fluid dynamics (CFD), which allowed visualization of the flow within the reactor and thus the effects of changes in the gas distribution configuration.

The silane-based FBR process is deceptively simple. The FBR itself is basically a vertical tube with a gas injector at the bottom, a top gas exit, a means to feed silicon seed into the top, and a "boot" or equivalent arrangement to withdraw the larger particles from the bottom. The entire reactor is held within an electrically heated furnace with careful multi-zone temperature control. The initial FBR used in REC's commercial production facility, the "A" model, was constructed of a single alloy material, but the second generation, the FBR-B, has a proprietary liner that dramatically reduces the metallic impurities [41]. Of course, the downstream material handling items can also contribute to the impurity burden as "surface metal" contamination. Combined with other downstream product handling improvements, REC's "FBR-B" has been specifically designed to produce electronic-quality silicon.

5.5 PROPERTIES OF GRANULAR SILICON AND HANDLING

Granular silicon produced in an FBR has the form of small spherical particles that vary in size from about 50 to 400 μm (see Figure 5.11).

A handling scheme of granular silicon is reported in Figure 5.12.

The mechanical properties of FBR silicon granules and of reference materials are reported in Table 5.4. These properties were measured by nanoindentation and microhardness tests for fracture toughness. The mean and standard deviation for between 16 and 28 tests for each sample is included in the table.

FIGURE 5.11 NEXT™ Si granular silicon. (REC Silicon, Inc., 2014. *REC Annual Report*, REC Silicon, Moses Lake, WA.)

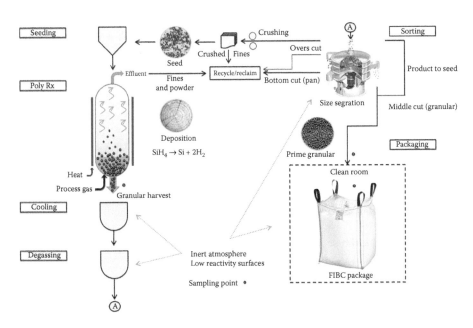

FIGURE 5.12 Handling scheme of granular silicon. (REC Silicon, Inc., 2014. *REC Annual Report*, REC Silicon, Moses Lake, WA.)

The chemical purity specifications of granular silicon, which have been established by the Semiconductor Equipment Manufacturers Institute (SEMI) in their document PV 017-0611 for solar-grade silicon are listed in Table 5.5.

One feature of granular FBR silicon is the presence of residual hydrogen. Hydrogen is trapped within the silicon granules during the silane pyrolysis and can "pop" when the granules are melted. To reduce this tendency, the granules can be annealed at high temperature after the granules are formed. Figure 5.13 shows the

TABLE 5.4
Mechanical Properties of Granular Silicon and of Reference Materials

Sample	Hardness (GPa)	Elastic Modulus (GPa)	Toughness (MPa m$^{1/2}$)
As grown granular	9.7 ± 0.4	164 ± 3	0.60 ± 0.05
Annealed	9.9 ± 0.8	149 ± 3	0.61 ± 0.13
(001) Single crystal	9.7 ± 0.3	146 ± 2	0.61 ± 0.13
Bulk Siemens polysilicon	10.0 ± 0.7	157 ± 3	0.73 ± 0.10

Source: Data of M. Zbib et al., 2009. *J. Mater. Sci.*, 45, 1560–1566.

TABLE 5.5
Bulk Chemical Properties of Granular Silicon by Grade

Grade	i	ii	iii	iv
Acceptors B, Al	<1 ppba	<20 ppba	<300 ppba	<1000 ppba
Donors P, As, Sb	<1 ppba	<20 ppba	<50 ppba	<720 ppba
Bulk O	ns	ns	ns	ns
C	<0.3 ppba	<2 ppba	<5 ppba	<100 ppba
Transition metals Ti, Cr, Fe, Ni, Cu, Zn, Mn	<10 ppba	<50 ppba	<100 ppba	<200 ppba
Alkali and alkali earth Na, K, Ca	<10 ppba	<50 ppba	<100 ppba	<200 ppba

Note: The total hydrogen and chlorine should be sufficiently low so that particles do not explode on heating. ns = not specified.

effects of annealing temperature on the residual hydrogen, indicating that temperature in excess of 600°C should be used to reduce significantly the content of hydrogen compounds.

5.6 HIGH-PRESSURE FBR

SunEdison, the successor to Ethyl Corp. and MEMC, using the knowledge and experience gained from nearly 30 years of experience with silane-based FBR, has up-scaled that process recently as a high-pressure FBR [43]. Based on the cited patent, the reactor should be capable of producing silicon at a rate in excess of 100 kg/m^2 of reactor cross section and operates at a pressure in excess of 4 bar. Today, although

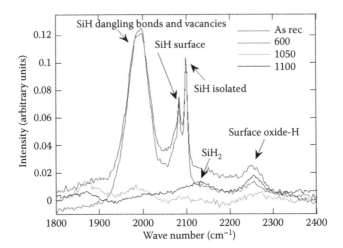

FIGURE 5.13 Hydrogen content of granular silicon. (Adapted from M. Zbib et al., 2009. *J. Mater. Sci.*, 45, 1560–1566.)

the start-up date of the delayed project was for the first quarter of 2015, no official documents are available to confirm or deny the statements made in the patent.

SiTec later announced that development work has commenced on a silane-based high-pressure FBR at their Seattle testing facility [44].

5.7 COMPARISON OF SILANE/FBR AND CONVENTIONAL SIEMENS

It is always instructive to compare the purity, quality, energy (CO_2 footprint), and cost of alternative process methods. These comparisons are somewhat difficult since the product form from the silane/FBR is granular and the conventional Siemens is a chunk or rod form factor. From a quality viewpoint, the conventional Siemens reactors have, so far, provided the highest quality. But even here, the silane precursor has the edge in quality over the TCS precursor, especially considering the critical electronically active impurities. Most importantly, the absence of corrosive chloride eliminates the need for a complex gas separation, purification, and recycle system when TCS is used.

The total energy for the silane/Siemens process relative to the TCS/Siemens route offers a closer comparison. The major cost elements in the purification and deposition processes can be summarized as follows:

1. Raw materials
   a. MG-Si: The cost of this critical raw material directly impacts the cash cost of the polysilicon product. The silicon component of cost can be minimized by minimizing the amount of "available" silicon lost in the consumption reactor effluent (either the hydrochlorination with HCl or the low-temperature "hydrogenation"). Control of MG-Si raw material

quality can have a significant impact. High levels of iron may promote better reaction kinetics, but iron silicides are not reactive and constitute a waste stream. The aluminum content of the MG-Si forms $AlCl_3$, which represents a loss of chloride.

b. Chloride replacement: Chloride-bearing materials originating from impurities in the MG-Si are lost as a fundamental part of the purification process. The process design can minimize the loss of chloride. "Purge" streams that remove impurities and associated chloride can have that loss replaced by importing HCl, STC, or even Cl_2 (to be combined with onsite H_2) whichever would be less expensive in the particular plant location.

c. Minimize other consumables: Consumables include ablative liners that reduce corrosion/erosion, carbon chucks in Siemens reactors, etc. Water is a precious commodity in certain geographical locations. Both the source and disposition of water resources can have a significant impact on the cash cost.

2. Energy

a. TCS/silane production: There are process alternatives that can reduce energy use in the processes that deliver these electronically pure materials. Opportunities are presented for minimizing energy in new plants that may not be cost effective to retrofit into older units. The use of modern process modeling programs permit consideration of alternative process configurations early in the design stage. Fortunately, the thermophysical properties of hydrochlorosilanes are well known and provide a sound basis for constructing a valid model. Energy recovery through process-to-process heat exchange, selection of operating conditions to minimize the need for deep refrigeration, or excessively high temperatures are all areas that can be explored to minimize the overall energy load.

b. Silicon deposition: This has traditionally been the process step that was the largest consumer of electric energy. The TCS/Siemens process energy consumption decreases as the capacity of the individual CVD reactors increases. Recently, GTAT has proposed a new reactor model that generates low-pressure steam with the claim that it reduces the net energy to less than 10 kWh/kg of Si versus the traditional designs that consume nearly 45 kWh/kg of Si [45]. GTAT also claims a net energy requirement in the TCS production of ~0.6 kWh/kg of silicon by efficient utilization of the low-pressure steam from Siemens reactors. Combining these concepts would forecast a net polysilicon energy utilization of about 10 kWh/kg [34].

c. Taking a global view of the entire process from MG-Si to final polysilicon by either the chlorosilane or silane route, the utilization of energy can be better appreciated. The TCS route must assume that the consumption of MG-Si is by the STC + H_2 + Si reaction to form TCS. Basically, 1 mol of Si is consumed (± the losses to impurities and minor process inefficiencies, which should be about the same by either the

TCS or silane route). Production of silane from TCS is a straightforward distillation/redistribution process, which is easily modeled. In the silane process, 3 mol of STC are generated for every mole of silane produced. In the TCS route, distillation is needed to refine the TCS and remove the electronically active impurities: mainly BCl_3 and PCl_3 (along with any low-molecular-weight hydrocarbons that find their way into the process). Those separations are more difficult than the separation of the bulk hydrochlorosilanes from silane and require taller columns operating at high reflux ratios: an energy-consuming operation. In the pyrolysis of TCS, the off-gases consist of a mixture of TCS, STC, HCl, and H_2. The H_2 must be separated from the HCl for recycle in the TCS pyrolysis reactor without introducing any harmful impurities and the TCS/STC mixture would be recycled to the distillation area where the STC is isolated for recycle to the MG-Si reactor. The total STC in that recycle loop is effectively the same as the amount recycled in the silane process (3 mol of STC/mole of silane).

But in the silane pyrolysis reactor, either FBR or Siemens, the only by-product is ultra-pure H_2, which is compressed and sent to the MG-Si consumption reactor. The low-pressure complicated separation of the chlorosilanes, H_2, and HCl is completely eliminated. In addition, the silane pyrolysis does not generate any quantity of chloropolysilane that is responsible for a loss of chloride and silicon value. The absence of chlorides in the elevated-temperature pyrolysis reactor using silane eliminates a significant corrosion issue and hence can yield a higher-purity polysilicon.

d. The CO_2 footprint for the production of polysilicon is dependent upon the geographical location of the plants and the process configuration. Fossil fuels are not inherently essential features of any of the production processes. Electric power is essential for the Siemens-type decomposer furnaces. The FBR reactors are favorably designed using electric heat, but the amount of energy required there is a small fraction of that required by the Siemens reactors. In the silicon consuming portion of the plant, gas- or oil-fired units can be effectively used to supply process heat to vaporize the chlorosilane feed to the FBR silicon consumption reactors to supply process energy for the distillation systems. Depending upon the availability and cost of hydroelectric power, those process heaters could be fueled with electricity. Such was the case for REC's first silane production facility at Moses Lake. That plant consequently had a very low carbon footprint. Initial capital cost usually favors fossil fuel for ordinary process heating. However, the life cycle cost could tip the balance toward electricity if power rates were favorable.

e. The cash cost for producing granular silicon via the silane process today is close to $10/kg. It has been steadily dropping as process optimization and efficiencies have been improving [26]. A detailed breakdown of the cost associated with the production of the high-purity gaseous

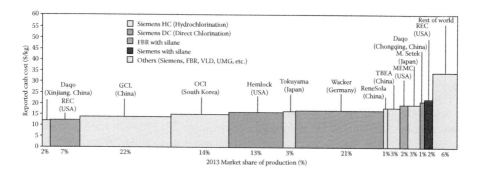

FIGURE 5.14 Supply curve of selected global polysilicon producers. (Adapted from R. Fu, T. James, and M. Woodhouse, 2015. *IEEE J. Photovoltaics*, 5, 518.)

silane intermediates (silane or TCS) is proprietary to all producers. The reported cash cost for various producers using different production technologies is presented in Figure 5.14. The "hydrochlorination" route defined by Fu as the reaction

$$2H_2 + 3SiCl_4 + Si \rightleftharpoons 4SiHCl_3 \qquad (5.9)$$

is favorably in the lowest-cost category, especially among the producers in China, who coincidently have the newest facilities. REC with its silane-based FBR process is in the lower level of this group, while REC's silane/Siemens process has costs higher than the bulk of major producers, regardless of the process. Hemlock and Wacker, using the direct chlorination process coupled with Siemens deposition furnaces, in spite of their considerable scale of operation, are not the lowest-cost producers today.

5.8 FUTURE OUTLOOK AND CONCLUSIONS

The growth in demand for granular silicon as a raw material for solar applications is expected to be strong. While MEMC (Pasadena) has been producing FBR silicon for several decades, it was not until REC demonstrated that it could also be produced at low cost did interest in granular silicon begin to pick up. The growth has been hindered by the high initial cost (silane) process of MEMC and by the lower quality of the flowable polysilicon relative to Czochralski silicon. Scale-up of the FBR process has not been easy.

It took REC from 1996 to 2009 to get the FBR-A version into commercial operation and the FBR-B is still not actualized commercially, although that technology will be used in the REC joint venture with Shaanxi Tian Hong in Yulin, China. But with a lower-cost structure and the improvements in product quality, which has been demonstrated over the past few years in pilot-scale operations, production quality levels are expected to increase. This increase in quality and consistency coupled with continuing cost reduction should lead to an increased use of granular polysilicon, although the Bernreuter's forecast of a 13% FBR silicon use in 2015 may probably not be achieved [47].

This should broaden the interest in flowable polysilicon to allow traditional Czochralski crystal growers to also benefit from its lower cost [27].

The technology of polysilicon will continue to advance with a focus on reduced cost of operation and capital [48]. As seen in Figure 5.14, the hydrochlorination step is dominant among the lowest-cost processes. Silane FBR provides an additional cost reduction. SiTec GmbH announced in September 2015 that they are developing a silane-based FBR that operates at elevated pressures up to about 20 bar while feeding 100% silane. They forecast an energy requirement in their mechanically mixed FBR of about 1 kWh/kg, and coupled with their advanced silane production design, they should achieve an overall power utilization of about 7 kWh/kg of granular polysilicon [44]. They expect the process to be developed for commercial implementation in about 2 years.

With the improved quality of FBR silicon, bulk crystallization to grow substrates for multi-crystalline solar cells should also be a major driver toward increased demand for and production of flowable polysilicon. The emphasis on cost will continue. The thrust on cost control would shift to focus on silane production since it appears to have the dominant low-cost position for new production plants. Cost reduction in silane production will focus on improving the efficiency of the chloride material balance, increased efficiency of utilization of the MG-Si, and optimizing energy use in the overall process. The energy to produce silane is now the largest energy cost component—a significant shift from legacy operations.

REFERENCES

1. W. T. Fairchild, 1964. Technology of silicon metal operation, *Proc. Elect. Furn. Conf.*, 21, 277–288.
2. B. Ceccaroli and S. Pizzini, 2012. *Processes in Advanced Silicon Materials for Photovoltaic Applications*, S. Pizzini (Ed.), John Wiley & Sons, Chichester, UK.
3. A. Schei, J. Tuset, and H. Tveit, 1998. *Production of High Silicon Alloys,* Tapir Forlag, Trondheim, Norway.
4. E. Marlett, 1986. Process for production of silane, US Patent 4,632,816.
5. J. Bossier, 1991. Process for preparation of silane, US Patent 5,075,092.
6. E. Marlett, 1988. Preparation of silane and amine alanes, US Patent 4,757,154.
7. R. Allen, 1989. US Patent 4,868,013.
8. L. Woerner and E. Moore, 1982. Process for production of polycrystalline silicon, US Patent 4318942.
9. L. Litz and S. Ring, 1964. US Patent 3,163,590.
10. P. Ehrlich, 1963. Alkaline earth metals, in *Handbook of Preparative Inorganic Chemistry*, 2nd ed. vol 1. G. Brauer, Ed., Academic Press, New York, p. 920.
11. H. Ishisaka, 1964. Apparatus for the manufacture of high purity elemental silicon by thermal decomposition of silane, US Patent 3,147,141.
12. W. Breneman and J. Mui, 1976. *A Process for High Volume Low Cost Production of Silane*, Project 486S80 File No. SVP-76-1, U.S. Department of Energy, Union Carbide Corp., Sistersville, WV.
13. E. Rochow, 1945. Preparation of organosilicon halides, US Patent 2,380,995.

14. D. L. Bailey, 1958. Disproportionation of chlorosilanes employing amine-type catalysts, US Patent 2,834,648.
15. H. Bradley, 1974. Production of silicon metal from dichlorosilane. US Patent 3,824,121.
16. J. C. Litteral, 1978. Disproportionation of chlorosilane, US Patent 4,113,845.
17. C. J. Bakay, 1976. Process for making silane, US Patent 3,986,199.
18. E. Christiansen, 1985. *Flat Plate Solar Array Project*, NASA Task RE-152, Amendment 66, DOE/NASA IAA No. DE-AIOI-76ET20356, Jet Propulsion Laboratory, Pasadena, CA.
19. G. Wagner and C. E. Erickson, 1952. Hydrogenation of halogensilanes, US Patent 2,595,620.
20. J. Mui and D. Seyferth, 1981. DOE/JPL 955382-79/8 (DE81025543), Massachusetts Institute of Technology, Cambridge, MA.
21. J. Mui, 1979. DOE/JPL 955382-79/8 (DE81025543), Massachusetts Institute of Technology, Cambridge, MA.
22. M. Zbib, 2011. Effect of solute hydrogen on toughness of feed stock polycrystalline silicon for solar cell applications, PhD thesis, Washington State University, Pullman, WA.
23. I. Röwer et al. 2002. The catalytic hydrogenation of chlorosilanes, in *Silicon for the Chemical Industry VI*, Loen, Norway, pp. 209–224.
24. H. Walter, G. Röwer, and K. Bohmhammel, 1996. Mechanism of the silicide-catalyzed hydrodehalogenation of silicon tetrachloride to trichlorosilane, *J. Chem. Soc. Farad. Trans.*, 92, 4605–4608.
25. W. Breneman, 1977. *Low Cost Silicon Solar Array Project*, Project 486S80 File No. SVP-77-1, Union Carbide, Sistersville, WV.
26. REC Silicon, Inc., 2014. *REC Annual Report*, REC Silicon, Moses Lake, WA.
27. C. Roselund, 2015. The slow grind of FBR polysilicon, *PV Mag.*, September, 88–91.
28. P. Gupta, 2014. Production of polycrystalline silicon in substantially closed-loop systems that involve disproportionation operations, US Patent 8,715,597.
29. GT Advancd Technologies, 2015. Generation 2 convertor, http://www.gtat.com/resources-product-sheets.htm
30. Advanced Silicon Materials, Inc., 2005. *Bulk Silane Specification*, Moses Lake, WA.
31. L. Britton, 1990. Combustion hazards of silane and its chlorides, *Plant Oper. Progr.*, 9, 16–38.
32. F. Taminini, 1997. Ignition and reactivity characteristics of releases of 100% silane. *Sematech Silane Conference*, Factory Mutual, Cambridge, MA.
33. R. B. Richardson, 1984. *Thermodynamic Package SGS-69*, SGS-69, Union Carbide Corp., Washougal, WA.
34. GTAT, 2015. GTAT silane production unit, http://www.gtat.com/products-and-service-silane-production-unit.ht (accessed September 7, 2015).
35. M. Osborne, 2015. http://www.pv-tech.org/news/rec_silicon_to_build_19000mt_polysilicon_plant_in_china_under_jv_partnershi (accessed November 25, 2015).
36. A. A. Onischuk, 2001. Mechanism of thermal decomposition of silanes, *Russ. Chem. Rev.*, 4, 321–332.
37. R. Elbert, 1983. CA Patent 1144739.
38. Roberts et al., 1980. *Development and Design for a Silion Powder Consolidation System*, JPL/DOE Contract 954334, Kayex Corp., Rochester.
39. F. Padovani, 1980. Silicon refinery, US Patent 421,397.
40. R. Allen, 1993. US Patent 5,242,671.
41. M. Miller, 2015. US Patent 2,015,017,787 A1.
42. M. Zbib, M. C. Tarun, M. G. Norton, D. F. Bahr, R. Nair, N. X. Randall, and E. W. Osborne, 2009. Mechanical properties of polycrystalline silicon solar cell feed stock grown via fluidized bed reactors, *J. Mater. Sci.*, 45, 1560–1566.

43. S. Bhusarapu, P. Gupta, and Y. Huang, 2013. Production of polycrystalline silicon by thermal decompositon of silane in a fluidized bed reactor, WO Patent 2013/049325.

44. SiTec GmbH announces new monosilane decomposition technology, http://www.centrotherm-pv.com/press/news/news-of-subsidiaries/news-tochterunternehmen/article/sitec-gmbh-announces-new-monosilane-decomposition-technology.html (accessed November 15, 2015).

45. GT AT. SDR 600 CVD reactor, http://www.gtat.com/Collateral/Documents/English-US/Polysilicon/SDR%E2%84%A2600%20CVD%20REACTOR.pdf (accessed November 14, 2015).

46. R. Fu, T. James, and M. Woodhouse, 2015. Economic measurements of polysilicon, *IEEE J. Photovoltaics*, 5, 518.

47. J. Bernreuter, 2014. Polysilicon market on a roller coaster ride, *Polysilicon Market Reports*, Bernreuter Research, Würzburg, Germany.

48. GT Advanced Technologies. 2015. *GT Advanced Technologies—Silane Production*, http://www.gtat.com/products-and-services-silane-production-unit.htm.

6 Thermodynamic Research for the Development of Solar Grade Silicon Refining Processes

Kasuki Morita

CONTENTS

6.1 THERMODYNAMIC PROPERTIES OF IMPURITIES IN MOLTEN SILICON

Although some practical development of the metallurgical refining processes for solar grade silicon feedstock was started in the early 1980s (see Chapters 2, 3, and 4), fundamental information on thermodynamic properties of impurities in molten silicon was not available until research about this topic was started in Japan. Suzuki et al. [1–3] clarified the possibility of phosphorus removal by vacuum treatment and boron removal by slag treatment and Ar–H_2O plasma oxidation processes, while

Ikeda and Maeda [4,5] succeeded in removing phosphorus by electron beam (EB) heating in vacuum conditions and boron by Ar–H_2O plasma.

On the basis of these fundamental results of the Sano and Maeda groups, within a NEDO* project addressed at the development of solar grade silicon (SOG-Si) production from MG-Si, Kawasaki Steel Corporation [6–8] (now JFE Steel Corporation) and some other companies, established the practice for both phosphorus and boron removal processes, although only few thermodynamic data of impurities, except those concerning C, N, and O in molten Si, were available at that moment.

Starting from the late 1990s, the Sano and Morita groups clarified such data for B [9], P [10], Ca [11,12], Al [11,12], Mg [11,12], Fe [13], and Ti [13], some of which are extremely important in the optimization of the metallurgical refining processes, as will be illustrated in the following sections.

6.1.1 BORON

The thermodynamic properties of boron in molten silicon were first investigated by the chemical equilibration technique using N_2 + Ar gas and boron nitride by Noguchi et al. [14] in 1994, and by Yoshikawa and Morita [9] a few years later, with the aim of suppressing the effect of supersaturation of Si_3N_4 during equilibrium measurements.

As the result of these studies, the activity coefficient γ_B of boron, where the activity of B, a_B is given by the product $\gamma_B x_B$ and x_B is the B concentration in molten silicon and its temperature dependence was obtained

$$\ln \gamma_B = 1.19(\pm 0.25) + 289(\pm 450)T^{-1} \tag{6.1}$$

by equilibrating molten Si alloys with BN and Si_3N_4.

In the NEDO process, the boron removal was performed by plasma treatment with the addition of water vapor to form HBO gas, but the oxidation removal by slag treatment was also found possible, due to considerably large activity coefficient of boron in molten silicon. Figure 6.1 displays the composition dependence of the partition ratio $L_D{}^\dagger$ of boron between molten silicon and CaO–SiO_2($-CaF_2$) slags, measured by Teixeira and Morita [15].

Although a single slag treatment may not be sufficient to obtain SOG-Si from MG-Si, the possibility of boron removal was, nevertheless, demonstrated.

Applying a different approach, based on the "borate capacity" $C_{BO_3^{3-}}$ of the slag

$$C_{BO_3^{3-}} = \frac{(\text{mass}\%BO_3^{3-})}{a_B p_{O_2}^{3/4}} \tag{6.2}$$

based on the oxidation reaction of boron in the slag

$$B_{Si} + \frac{3}{4}O_{2(g)} + \frac{3}{2}O_{slag}^{2-} \rightleftharpoons (BO_3^{3-})_{slag} \tag{6.3}$$

---

* New Energy and Industrial Technology Development Organization.
† Partition ratio and segregation coefficient are equivalent.

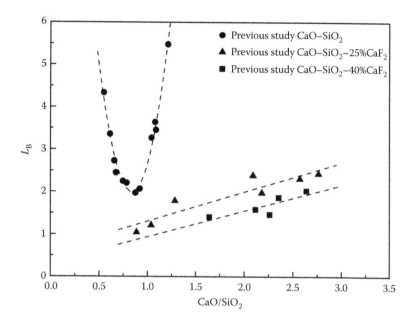

FIGURE 6.1   Boron partition ratio between a CaO–SiO$_2$ slag and the silicon phase as a function of the final basicity of the slag equilibrated at 1823 K. (Reproduced with permission of the Japan Institute of Metals and Materials after L. A. V. Teixeira and K. Morita, 2009. Removal of boron from molten silicon using CaO–SiO$_2$ based slags, *ISIJ Int.*, 49, 783–787.)

the boron partition ratio L_B between the slag and the molten silicon phase can be predicted at a certain oxygen partial pressure, p$_{O_2}$, using the activity coefficients of boron in molten silicon.

6.1.2   Phosphorus

The thermodynamic property of phosphorus in molten silicon was originally investigated by Miki et al. [10] using the transpiration method. It works by equilibrating a molten Si–P alloy with P at a controlled phosphorus partial pressure (P$_{P_2}$ = 0.00859 − 0.492 [Pa]) in the temperature range 1723–1848 K with Ar gas saturated with red phosphorus at a fixed temperature. At the end of the experiment, the phosphorus analysis in the silicon phase is carried out.

As the result of these measurements, the Gibbs free energy of the dissolution reaction of monoatomic phosphorus into molten silicon

$$\frac{1}{2} P_2^v \rightleftharpoons P_{Si} \tag{6.4}$$

$$\Delta G^o = -139(\pm 2) + 0.0434(\pm 0.0101)T \text{ (kJ/mol)} \tag{6.5}$$

was obtained.

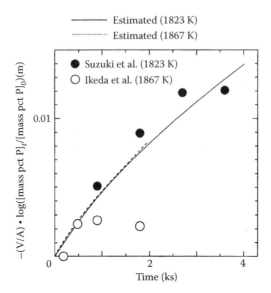

FIGURE 6.2 Relationship between time for vacuum treatment and $-(V/A) \log([\text{mass}\%$ $P]/[\text{mass}\% \ P]_0)$. (After T. Miki, K. Morita, and N. Sano, 1996. Thermodynamics of phosphorus in molten silicon, *Metall. Mater. Trans. B*, 27B, 937–942. Reproduced with permission of TMS.)

From these experimental results, under the condition that the phosphorus content in silicon is lower than 50 ppma, monoatomic P was found to be the dominant vapor species rather than the diatomic P_2 species.

It is interesting to mention here that the results of the process developed in the NEDO project, addressed at the electron beam (EB) refining of MG-Si in vacuum conditions, could be rationalized by supposing that the dephosphorization process at 1823 K was controlled by the vaporization of monoatomic phosphorus species from the surface of molten silicon, as shown in Figure 6.2, where V is the volume of the silicon sample, A is its surface area, P_t is the P mass at the time t, and P_0 is the P mass at the time 0.

6.1.3 ALUMINUM, MAGNESIUM, AND CALCIUM

Aluminum, magnesium, and calcium are the elements which are easier to be oxidized as compared with silicon, and have also relatively high vapor pressures. It could be, therefore, expected that these impurities dissolved in molten silicon could be removed by oxidation or vacuum refining.

Hence, the thermodynamic properties of such elements were measured by equilibrating a molten silicon alloy with solid or liquid oxides. To this scope, a Si–Al alloy saturated with Al_2O_3 at 1723–1848 K was used for aluminum, a Si–Mg alloy in equilibrium with a $MgO-SiO_2-Al_2O_3$ slag doubly saturated with $MgSiO_3$ and SiO_2 at 1698–1798 K was used for magnesium, and a Si–Ca alloy in equilibrium with a SiO_2 saturated $CaO-SiO_2$ slag at 1723–1823 K was used for calcium [11].

By fixing the chemical potentials (activities) of constituent oxides, the activity of the element can be derived from the Gibbs free energy for the formation of the oxides

TABLE 6.1

Activity Coefficients of Aluminum, Magnesium, and Calcium in Molten Silicon at Infinite Dilution Relative to Pure Liquid Substances and Self-Interaction Parameters in Molten Silicon

Element	Activity Coefficient (γ)	Temperature Range (K)	Interaction Parameters (ε)	Temperature Range (K)
Al	$\ln \gamma(Al) = -3.610\,T^{-1} + 0.452$	1723–1848	$\varepsilon_{Al}^{Al} = 1.0010^5\,T^{-1} - 40.1$	1723–1848
Mg	$\ln \gamma(Mg) = -11.300T^{-1} + 4.51$	1698–1798	$\varepsilon_{Mg}^{Mg} = 6.02$	1723
Ca	$\ln \gamma(Ca) = -14.300T^{-1} + 1.55$	1723–1823	$\varepsilon_{Ca}^{Ca} = 55.600\,T^{-1} - 22.1$	1723–1823

Source: Adapted from T. Miki, K. Morita, and N. Sano, 1999. *Mater. Trans. JIM*, 40, 1108–1116.

(SiO_2, Al_2O_3, MgO, and CaO) assuming that silicon follows Raoult's law in each molten alloy. In addition to equilibrium experiments, the effect of self-interaction of aluminum and calcium in molten silicon was investigated by the Knudsen effusion method [12]. The results obtained in terms of activity coefficients at infinite dilution and the first-order self-interaction parameters* are summarized in Table 6.1.

It could be seen from Figure 6.3 that the vaporization of Al, Mg, and Ca from molten silicon works faster than for phosphorus. The same figure also shows the key role of the surface area of the sample on the yield of the process.

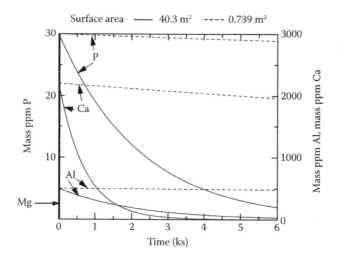

FIGURE 6.3 Relationship between vacuum treatment time and impurity contents of silicon at 1823 K. (After T. Miki, 1999. Thermodynamic property of impurity elements in molten silicon, Doctoral Thesis, The University of Tokyo, Tokyo, Japan.)

---

* The interaction parameters are used to account for the mixing enthalpy of nonideal solutions [41].

6.1.4 TITANIUM AND IRON

Among the lifetime killer elements in solar cell silicon (see Chapter 1), the thermodynamic properties of titanium and iron in molten silicon were studied in full details. Although their segregation coefficients are considerably small, a single directional solidification (DS) or a Czochralski (CZ) growth is not sufficient to reach the required concentration levels for high efficiency solar cells, and additional processes should be foreseen, in alternative to a too expensive, double crystallization process. In order to evaluate the possibility of use of alternative physical or chemical purification processes, the thermodynamic properties of these elements in molten silicon were investigated [13]. Molten silicon–titanium alloys and molten silicon–iron alloys were equilibrated with molten lead at 1723 K, which has a limited mutual solubility with both alloys. From the experimental values of the distribution ratio of titanium and iron between the two metallic phases and from the (known) thermodynamic properties of titanium and iron in molten lead, the activity coefficients of titanium and iron in molten silicon at infinite dilution relative to pure liquid titanium and iron$_F$, $\gamma^o_{Ti\ in\ Si}$ and $\gamma^o_{Fe\ in\ Si}$, at 1723 K could be obtained, which are 4.48×10^{-4} and 2.85×10^{-2}, respectively.

Also, the self-interaction parameters of titanium and iron in molten silicon, ε^{Ti}_{Ti} and ε^{Fe}_{Fe} were determined and shown to hold 3.97 and 3.17, respectively, at 1723 K. For iron in molten silicon, the temperature dependence of activity coefficients and of the self-interaction parameters was determined by the Knudsen effusion method [17]

$$\ln \gamma_{Fe} = 4.10 - \frac{13.2 \times 10^3}{T} \tag{6.6}$$

$$\varepsilon^{Fe}_{Fe} = \frac{75.6 \times 10^3}{T} - 40.7 \tag{6.7}$$

Although pure titanium is easier to be oxidized compared to pure silicon, the activity coefficient of titanium in molten silicon is too small to allow Ti to be removed from molten silicon by oxidation refining. The same condition holds for iron that cannot be removed from molten silicon by the oxidation process nor by vacuum treatment. Even chlorination refining (see Chapter 3), forming volatile titanium or iron chlorides, was found to be hopeless due to the high vapor pressure of silicon chlorides.

Hence, the acid leaching of Si–Ca alloys was thought to be one of the preliminary treatments which could be carried out for a preliminary purification (see also Chapters 3 and 4), and Si–Ca–Fe alloys were subjected to an acid leaching with aqua regia [18].

It was demonstrated that only in the case of grain boundary segregation of an acid-soluble $CaSi_2$ phase, as shown in Figure 6.4, the removal of the iron-rich eutectic phase was possible. This conclusion was confirmed noting that the amount of calcium in the Si–Ca alloy should be high enough so that the precipitated phase should be $CaSi_2$, with $FeSi_2$ inclusions. Otherwise, as shown in Figure 6.5, if the presence of isolated precipitates of $FeSi_2$ and $CaSi_2$ does occur, the $FeSi_2$ precipitates will remain undissolved after the leaching process. Figure 6.6 shows, eventually, the influence of the Ca/Fe ratios on the removal ratio of Fe by aqua regia leaching at 368 K for 3.6 ks, that reaches at a maximum above $x_{Ca}/x_{Fe} > 5$.

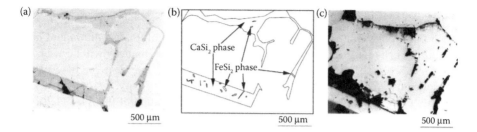

FIGURE 6.4 (a) Optical image of Si–Ca–Fe alloy. (Si–8.64%Ca–0.756%Fe, before acid leaching). (b) Microstructure of Si–Ca–Fe alloy. (c) Optical image of Si–Ca–Fe alloy. (Si–8.64%Ca–0.756%Fe, 5 min in aqua regia). (Reprinted from the *Intermetallics*, 11, K. Morita and T. Miki, Thermodynamics on solar-grade-silicon refining, 1111–1117. Copyright 2003, with permission from Elsevier.)

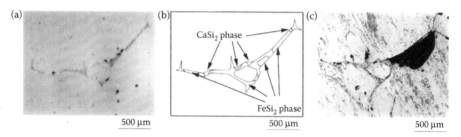

FIGURE 6.5 (a) Optical image of Si–Ca–Fe alloy. (Si–0.929%Ca–1.21%Fe, before acid leaching). (b) Microstructure of Si–Ca–Fe alloy. (c) Optical image of Si–Ca–Fe alloy. (Si–0.929%Ca–1.21%Fe, 5 min in aqua regia). (Reprinted from the *Intermetallics*, 11, K. Morita and T. Miki, Thermodynamics on solar-grade-silicon refining, 1111–1117. Copyright 2003, with permission from Elsevier.)

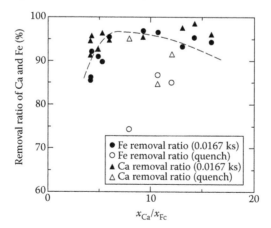

FIGURE 6.6 Relationship between the ratio of calcium to iron content in the alloy and the removal ratio of calcium and iron. (Reproduced with permission of the Japan Institute of Metals and Materials after T. Sakata, T. Miki, and K. Morita, 2002. Removal of iron and titanium in poly-crystalline silicon by acid leaching, *J. Jpn. Inst. Met.*, 66, 459–465.)

6.1.5 INTERACTION PARAMETERS

When a third element j is added to molten silicon containing a small amount of a specific impurity i, the thermodynamic property of the first impurity will be affected by i–j interactions. The quantitative effects of such interaction are given in terms of interaction parameters, which measure the changes in the activity coefficients of the impurity i in molten silicon as a function of the concentration of the third element. The first-order interaction parameter[*] is defined by Equation 6.8 [20],

$$\varepsilon_i^j = \left(\frac{\partial \ln \gamma_i}{\partial x_j} \right)_{x_i = x_j = 0} \tag{6.8}$$

Higher order parameters should be introduced for concentrated solutions.

The interactions between calcium and boron [21], and calcium and phosphorus [22] in molten silicon were investigated, in order to speculate about the yield of P and B removal from Ca–Si alloys using an acid leaching treatment. The Si–Ca alloy was equilibrated with Si_3N_4 and hex-BN for boron, and molten lead phosphorus-saturated for phosphorus, and values of the interaction parameters between calcium and boron, and calcium and phosphorus in molten silicon at 1723 K were determined to hold:

$$\varepsilon_{Ca}^B = \varepsilon_B^{Ca} = -3.08(\pm 0.84) \tag{6.9}$$

$$\varepsilon_{Ca}^P = \varepsilon_P^{Ca} = -14.6(\pm 1.7) \tag{6.10}$$

While the interaction between calcium and boron is rather small, that between Ca and P shows a larger negative value. Therefore, a larger effect of the calcium addition to silicon on the phosphorus removal could be expected, as experimentally demonstrated in Figure 6.7, which shows also the fairly good agreement of the experimental results with the calculated ones from the interaction parameters.

6.2 SOLVENT REFINING

Considering that the solubility of major impurities in solid silicon decreases with the decrease of the temperature (see Figure 6.8 [23]), a low temperature solvent refining process could be profitably envisaged, using a Si–Al alloy as the solvent [24–31]. The process (see also Chapter 3) is carried out (a) by alloying MG-Si with aluminum to form the Si–Al solvent, (b) followed by the partial solidification of silicon from the solvent at around 1273 K, and (c) by collection of the refined silicon crystals by acid leaching.

The high merits of this process, which is now developed at the industrial scale (see Chapter 3), depend on the extremely small segregation coefficients of impurities at a temperature lower than the melting point of silicon and were experimentally confirmed, as can be seen in the next sections.

---

[*] At infinite dilution.

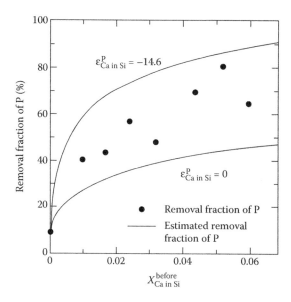

FIGURE 6.7 Comparison between estimated removal fraction of phosphorus in silicon and experimental results for acid leaching treatment as a function of calcium content in the Si–Ca–P alloy. (With kind permission from Springer Science+Business Media: *Metall. Mater. Trans.*, Thermodynamic study of the effect of calcium on removal of phosphorus from silicon by acid leaching treatment, 35B, 2004, 277–284, T. Shimpo, T. Yoshikawa, and K. Morita.)

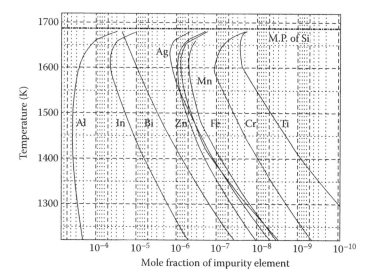

FIGURE 6.8 Solid solubility of impurity elements in silicon. (After F. A. Trumbore, 1960. Solid solubilities of impurity elements in germanium and silicon, *Bell Syst. Tech. J.*, 39, 206–233.)

6.2.1 Thermodynamic Principles

During the partial solidification of silicon from a Si–Al alloy, the impurity elements are subject to a solid/liquid segregation process, where the segregation ratio, k_i, is defined as the ratio, in mole fraction, of the concentration of the impurity i in solid silicon and in the solvent

$$k_i = \frac{x_i^{\text{Si}}}{x_i^{\text{Si–Al}}} \tag{6.11}$$

The segregation ratios of phosphorus (1173–1373 K) and boron (1273–1473 K) were measured by the temperature gradient zone melting method [25,28]. An Al–P or Al–B foil was initially located between two single crystalline silicon plates, and placed in the temperature gradient region of a horizontal resistance furnace. By analyzing the phosphorus or boron concentration in solid silicon and in the melt zone after quenching the sample, the segregation ratios could be determined, and shown quite slower than those measured at the melting temperature of silicon, in good agreement with the expected lower solubility shown in Figure 6.8. Thus, the removal of phosphorus and boron by the partial solidification of silicon from a Si–Al solvent could be expected to be effective.

A problem, instead, arises when considering the behavior of metallic impurities, due to the difficulty in measuring the segregation ratios between solid silicon and a Si–Al solvent because of their extremely small solubilities in solid silicon.

Therefore, the segregation behavior of metallic impurities was thermodynamically evaluated by using [26] the following equation:

$$\ln k_i = \ln \frac{x_i^{\text{Si}}}{x_i^{\text{Al–Si}}} = -\frac{\Delta G_i^{\text{fus}} + \overline{\Delta G_i}^{\text{ex,Al–Si}} - \overline{\Delta G_i}^{\text{ex,Si}}}{RT} \tag{6.12}$$

Here, ΔG_i^{fus}, $\overline{\Delta G_i}^{\text{ex,Al–Si}}$, and $\overline{\Delta G_i}^{\text{ex,Si}}$ are, respectively, the Gibbs free energy of fusion of the impurity i, the excess Gibbs free energy of the impurity element in a Si–Al solvent, and that of i in solid silicon, respectively. The calculated segregation ratios obtained are reported in Table 6.2 and found to be much smaller than the segregation coefficients between solid and liquid silicon at its melting point. In addition, their values decrease as the temperature decreases.

This theoretical modeling could be used to calculate the refining yield when a MG-Si sample with Fe:2000 ppmw, Ti:100 ppmw, P:20 ppmw, and B:20 ppmw is alloyed with pure aluminum to get a Si–55.3at%Al alloy, and then partially solidified at 1273 K. The estimated impurity contents of silicon after both solidification refining are shown in Figure 6.9 and compared with allowable contents[*] for SOG-Si [26].

Iron and titanium can be reduced below the allowable contents by the low temperature solidification, whereas the purification yield concerning phosphorus and boron are inadequate for SOG-Si.

---

[*] The allowable impurity content follows the Yoshikawa and Morita definition in Reference 26 that may differ from the solar silicon definition discussed in Chapter 1.

TABLE 6.2

Calculated Segregation Ratios of Metallic Impurities between Solid Silicon and Si–Al Solvent at 1073 and 1273 K with Segregation Coefficients at the Melting Point of Silicon

Element	Segregation Ratio in Si/Al Solvent ($T = 1073$ K)	Segregation Ratio in Si/Al Solvent ($T = 1273$ K)	Segregation Coefficient at Silicon Melting Point
Fe	1.7×10^{-11}	5.9×10^{-9}	6.4×10^{-6}
Ti	3.8×10^{-9}	1.6×10^{-7}	2.0×10^{-6}
Cr	4.9×10^{-10}	2.5×10^{-8}	1.1×10^{-5}
Mn	3.4×10^{-10}	4.5×10^{-8}	1.3×10^{-5}
Ni	1.3×10^{-9}	1.6×10^{-7}	1.3×10^{-4}
Cu	9.2×10^{-8}	4.4×10^{-6}	4.0×10^{-4}
Zn	2.2×10^{-9}	1.2×10^{-7}	1.0×10^{-5}
Ga	2.1×10^{-4}	8.9×10^{-4}	8.0×10^{-3}
In	1.1×10^{-5}	4.9×10^{-5}	4.0×10^{-4}
Sb	3.4×10^{-3}	3.7×10^{-3}	2.3×10^{-2}
Pb	9.7×10^{-5}	2.9×10^{-4}	2.0×10^{-3}
Bi	1.3×10^{-6}	2.1×10^{-5}	7.0×10^{-4}

Source: Adapted from T. Yoshikawa and K. Morita, 2003. *Sci. Technol. Adv. Mater.*, 4, 531–537.

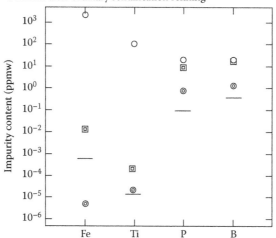

FIGURE 6.9 Comparison of impurity concentration of silicon after a low temperature solidification refining of MG–Si with an Si–Al solvent at 1273 K and that after an ordinary silicon solidification refining. (After T. Yoshikawa and K. Morita, 2005. Thermodynamics on the solidification refining of Si with Si–Al melts, in *Proceedings of the EPD Congress 2005, TMS Annual Meeting*, San Francisco, California, 549–558. Reproduced with permission of TMS.)

6.2.2 Si–Al Solvent Refining Test with Induction Heating

Test refining was carried out under induction heating that allows an intimate mixing of the charge. When a Si–Al alloy is melted and cooled down under induction heating, solidified silicon is agglomerated to either the top or bottom as shown in Figure 6.10 [27]. This solidification allows us to collect silicon crystals selectively from the alloy and to considerably reduce the loss of aluminum and silicon during the acid leaching process.

Synthetic samples of MG-Si with the composition shown in Table 6.3 were alloyed with pure aluminum to prepare a Si–55.3at%Al alloy for the tests S-01 and S-02 and a Si–64.6at%Al alloy for the tests S-03 and S-04. Liquidus temperatures of the alloys are 1273 and 1173 K, respectively. After heating to and holding at 1323 K for S-01 and S-02 and at 1223 K for S-03 and S-04 under induction heating, the alloys were cooled to 873 K. The refined silicon was collected after acid leaching of the particles with aqua regia and subjected to the chemical analysis. Impurity contents are summarized in Table 6.3 [27].

The removal fraction of boron was larger than expected, due to the TiB_2 formation with soluble titanium in the alloy [28,29]. The removal fraction of phosphorus agreed, instead, fairly well with calculated values (solidified from 1273 K; removal fraction 94.7%) [25].

In spite of the overall high purification yield achieved by the partial solidification of silicon from a Si–Al solvent under induction heating, the aluminum content is

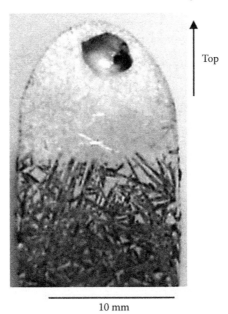

Top

10 mm

FIGURE 6.10 Cross-section of a solidified Si–55.3 mol%Al alloy under induction heating. (Reproduced with permission of the Japan Institute of Metals and Materials after T. Yoshikawa and K. Morita, 2005. Refining of Si by the solidification of Si-Al melt with electromagnetic force, *ISIJ Int.*, 45, 967–971.)

TABLE 6.3
Impurity Contents of Refined Silicon and Impurities Removal Fractions after Test Refining

Impurity (ppmw)	Mg–Si	S-01	S-02	S-03	S-04	Removal Fraction (%)	S-01	S-02	S-03	S-04
Fe	4500	13	13	47	36	Fe	99.7	99.7	99.0	99.2
Ti	691	52	2.7	7.7	5.6	Ti	99.2	99.6	98.9	99.2
Al	1280	599	534	538	453	Al	53.1	58.1	57.8	64.5
B	56	0.81	0.88	0.98	0.99	B	98.6	98.4	93.3	98.2
P	36	0.93		0.42	0.66	P	97.4	96.7	98.8	98.2

Source: Adapted from T. Yoshikawa and K. Morita, 2005. *ISIJ Int.*, 45, 967–971.

FIGURE 6.11   Microstructure of the bottom of the solidified Si–55.3 mol%Al alloy in the test refining. (Reprinted from the *Trans. Nonferr. Met. Soc. China*, 21, K. Morita and T. Yoshikawa, Thermodynamic evaluation of new metallurgical refining processes for SOG-silicon production, 685–690, Copyright 2011, with permission from Elsevier.)

larger than its equilibrium solubility in silicon at the experimental temperature [24] (see phase diagrams of Al–Si system in Figure 3.20), due to the entrapped solvent in silicon as shown in Figure 6.11 [32].

In order to prevent the solvent entrapment during the solidification, further work on silicon crystal growth is still under way.

6.2.3   SILICON CRYSTAL GROWTH IN A Si–Al SOLVENT BY DIRECTIONAL SOLIDIFICATION

Directional solidification (DS) was carried out to obtain bulk silicon crystals from a Si–55.3at%Al alloy [31]. The sample was cooled from 1323 to 1173 K with a lowering rate of 0.02–0.08 mm/min in a temperature gradient of 1.5–4.0 K/mm. Figure 6.12 shows the vertical cross sections of the samples obtained at various cooling rates and temperature gradients.

As the cooling rate decreases, the interface of the silicon crystals at the bottom of the sample becomes flatter for both temperature gradients. The solid fractions of samples (c) and (f), in which the crystal–melt interface is flat, could be roughly estimated from the phase diagram of the Si–Al binary system [33]. Under the assumption

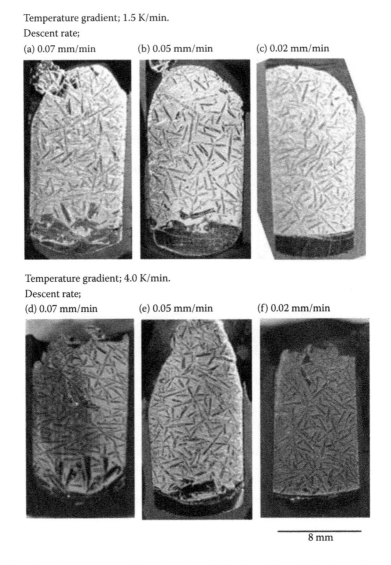

Temperature gradient; 1.5 K/min.
Descent rate;
(a) 0.07 mm/min   (b) 0.05 mm/min   (c) 0.02 mm/min

Temperature gradient; 4.0 K/min.
Descent rate;
(d) 0.07 mm/min   (e) 0.05 mm/min   (f) 0.02 mm/min

8 mm

FIGURE 6.12 Cross sections of directionally solidified Si–55.3 mol%Al alloys. (Reproduced with permission of the Japan Institute of Metals and Materials after Y. Nishi, Y. Kang, and K. Morita, 2010. Control of Si crystal growth during solidification of Si-Al melt, *Mater. Trans.*, 51, 1227–1230.)

that silicon crystal growth is diffusion controlled, the growth rate v, is given by the following equation, on the basis of the steady-state diffusion equation for silicon [26]:

$$v = D_{Si}^{Al-Si} \frac{\partial \xi_{Si}^{Al-Si}}{\partial x} \qquad (6.13)$$

where D_{Si}^{Al-Si}, ξ_{Si}^{Al-Si}, and x are the diffusion coefficient of silicon in the Al–Si alloy, the silicon concentration at the growth interface, and the distance in the crystal growth direction, respectively.

When the difference between the molar volumes of solid silicon and the Si–Al solvent is ignored, the growth rate of silicon could be rewritten

$$v = D_{Si}^{Al-Si} \frac{\partial \xi_{Si}^{Al-Si}}{\partial T} \frac{\partial T}{dx} \qquad (6.14)$$

as a function of the temperature gradient, $\partial T/\partial x$.

Since the maximum value of $\partial \xi_{Si}^{Al-Si}/\partial T$ is the liquidus slope at the growth temperature (see Figure 3.20), corresponding to temperature at the growth interface, the growth rates of the silicon crystals at 1273 K are calculated to be 1.5×10^{-3} and 4.0×10^{-3} mm/min at the temperature gradients of 1.5 and 4.0 K/mm, respectively. These calculated growth rates correspond well to faceted growth rates of the samples (c) and (f) estimated from Figure 6.12 as 8.7×10^{-4} and 2.3×10^{-3} mm/min, respectively. This indicates that the (faceted) growth of silicon from the Si–Al solvent is controlled by the diffusion of silicon and that silicon diffusion in the melt must be thus accelerated to achieve high efficiency crystal growth of silicon.

6.2.4 SOLVENT PURIFICATION AND COMBINED PURIFICATION PROCESSES

In solvent refining processes, the removal of the unwanted impurities in the solvent before and during solidification is desirable. The B removal is a particularly important example, since it would contaminate the silicon crystals thereafter produced.

The addition of titanium, in spite of being one of the worst lifetime killer elements, is very beneficial since it forms with B the compound TiB_2, stable at low temperatures and virtually insoluble in the solvent. When 300–1000 ppma Ti was added to a Si–Al solvent (B content: 100–200 ppma), the boron content of the refined silicon crystals after solidification and acid leaching was below 4 ppma, indicating that the immobilization of Ti as TiB_2 inclusions prevents silicon contamination produced in the solvent refining process [29].

Combining a slag treatment process with solvent refining may improve the ultimate refining ability, depending, however, on the reductive or oxidative properties of an alloy component, that is, on its Gibbs free energy of formation of the corresponding oxide (see Chapter 3, Section 3.2.2 and Figure 3.7). As an example, since aluminum interacts reductively with all the oxides of impurities more noble than Al (see Figure 3.7), the Si–Al solvent process is unsuitable in combination with an oxidative slag refining, such as that used for boron removal.

As another example, the $CaO-SiO_2-CaF_2$ slag treatment was applied in combination with a refining process using a Si–Sn alloy as the solvent. The aim was to investigate the thermodynamic stability of tin as a metal in the presence of the silicate slag [34] and its effect on the boron removal.

The distribution ratio of boron between the slag and the solvent was 160 at a solvent composition of Si–70 mol%Sn, much larger than the usual distribution ratio (up to 3) between the same slag and molten silicon [15].

The reaction for boron elimination by slagging is based on the following redox reaction:

$$B + \frac{3}{4}SiO_2 \rightleftharpoons BO_{1.5} + \frac{3}{4}Si \qquad (6.15)$$

In turn, the distribution ratio of B, on a molar basis, between the slag and silicon is given by the following equation:

$$L_B = \frac{\chi_{BO_{1.5}}}{\chi_B} = \frac{K \cdot (a_{SiO_2})^{3/4}(\gamma_B)}{(\gamma_{BO_{1.5}})(a_{Si})^{3/4}} \qquad (6.16)$$

The effect of silicon alloying with tin is double. First, in an alloy, the activity of silicon decreases and the distribution ratio increases. Furthermore, due to drastic decrease of the thermodynamic stability of boron in the presence of tin, and of the corresponding increase of the activity coefficient of boron in Si–Sn melts by three orders of magnitude, as shown in Figure 6.13 [35], the distribution ratio would be expected to significantly increase with the addition of tin.

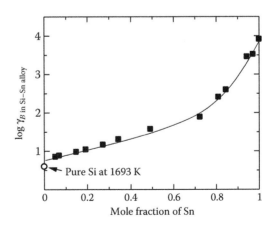

FIGURE 6.13 Activity coefficient of B in Si–Sn saturated with B or SiB_6 at 1673 K. (Reprinted from the *J. Alloys Compd.*, 529, X. Ma and T. Yoshikawa, Phase relations and thermodynamic property of boron in the silicon–tin melt at 1673 K, 12–16, Copyright 2012, with permission from Elsevier.)

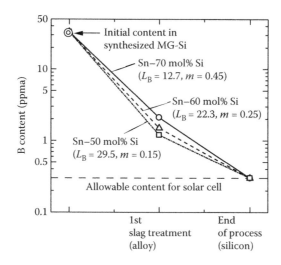

FIGURE 6.14 Change in B content during double slag treatment and partial solidification. The slag composition was 40.5%CaO–35.5%SiO$_2$–24 mol%CaF$_2$. L_B is the distribution ratio of B (slag/alloy, by weight) and m is the mass ratio of slag to alloy. (After X. Ma, T. Yoshikawa, and K. Morita, unpublished.)

In addition, a sequential process of double slag treatments followed by partial solidification using a Si–(50–70 mol%)Sn solvent was also examined [36]. The slag/alloy mass ratio was suitably changed in order to bring the distribution ratio of boron to a value suitable to reduce the boron content to an allowable level (0.3 ppma), as shown in Figure 6.14. It could be observed that a larger distribution ratio of boron is obtained by increasing the tin concentration in the alloy, with the advantage to also decrease the slag to silicon alloy ratio and the corresponding slag consumption.

6.2.5 Solvent with a High Vapor Pressure Component

As already emphasized in Chapter 3, Section 3.3.2.1, the separation of silicon crystals from the solvent is one of the key issues in the solvent refining process, and one possibility for its easier separation might be the use of a high-pressure alloy component which could be readily removed by evaporation. Morito et al. [37] investigated the possibility to segregate silicon from a Na–Si solvent by vaporizing the sodium. In this process, due to the small solubility of metallic impurities in the solvent, most of the impurities segregate as separate phases. If segregation occurs without heterogeneous mixing with the solidified silicon, the purified silicon can be easily removed.

The metallothermic reduction of SiCl$_4$ with Zn is another example of solvent refining with high-pressure components that has a long history starting around 1950 [38,39]. In this process, gaseous or liquid zinc have been used as major reductants for SiCl$_4$. SiCl$_4$ is generally of high purity, but if the reduction provides a liquid alloy (Zn–Si), from which a further purification of silicon is expected from solid/liquid segregation [40] and further removal of zinc, the alloy residue can be removed by evaporation, similar to the sodium evaporation mentioned earlier. Because of this advantage, zincothermic reduction has been regarded as a promising way to produce SOG-Si, although its reduction efficiency and the cyclic use of zinc must still be improved.

6.3 CONCLUSIONS

The design and implementation of silicon refining processes based on the use of slags or metallic solvents requires an appropriate knowledge of the thermodynamic properties of the solutions and of the solution components that could then be used for the development of thermodynamic models of accurate predictability.

The thermodynamic approach is shown also to be very useful for the design and/ or the optimization of dual refining processes, where a possible redox interference occurs among the different reactants.

The experimental determination of these properties and their application to a number of solvent systems has been the main focus of the research work carried out by our group and the key to the success of conventional and no conventional solvent refining processes, addressed at the synthesis of solar grade silicon.

REFERENCES

1. K. Suzuki, K. Sakaguchi, T. Nakagiri, and N. Sano, 1990. Gaseous removal of phosphorus and boron from molten silicon, *J. Jpn. Inst. Met.*, 54, 161–167.
2. K. Suzuki, T. Sugiyama, K. Takano, and N. Sano, 1990. Thermodynamics for removal of boron from metallurgical silicon by flux treatment, *J. Jpn. Inst. Met.*, 54, 168–172.
3. K. Suzuki, T. Kumagai, and N. Sano, 1992. Removal of boron from metallurgical-grade silicon by applying the plasma treatment, *ISIJ Int.*, 32, 630–634.
4. T. Ikeda and M. Maeda, 1996. Elimination of boron in molten silicon by reactive rotating plasma arc melting, *Mater. Trans. JIM*, 37, 983–987.
5. T. Ikeda and M. Maeda, 1992. Purification of metallurgical silicon for solar-grade silicon by electron beam button melting, *ISIJ Int.*, 32, 635–642.
6. H. Baba, N. Yuge, Y. Sakaguchi, M. Fukai, F. Aratani, and Y. Habu, 1991. Removal of boron from molten silicon by argon-plasma mixed with water vapor, in *Proceedings of the 10th EC Photovoltaic Solar Energy Conference*, Lisbon, Portugal, pp. 286–289.
7. N. Yuge, K. Hanazawa, K. Nishikawa, and H. Terashima, 1997. Removal of phosphorus, aluminum and calcium by evaporation in molten silicon, *J. Jpn. Inst. Met.*, 61, 1086–1093.
8. N. Nakamura, H. Baba, Y. Sakaguchi, and Y. Kato, 2004. Boron removal in molten silicon by a steam-added plasma melting method, *Mater. Trans.*, 45, 858–864.
9. T. Yoshikawa and K. Morita, 2005. Thermodynamic property of B in molten Si and phase relations for the Si–Al–B system, *Mater. Trans.*, 46, 1335–1340.
10. T. Miki, K. Morita, and N. Sano, 1996. Thermodynamics of phosphorus in molten silicon, *Metall. Mater. Trans. B*, 27B, 937–942.
11. T. Miki, K. Morita, and N. Sano, 1998. Thermodynamic properties of aluminum, magnesium and calcium in molten silicon, *Metall. Mater. Trans. B*, 29B, 1043–1049.
12. T. Miki, K. Morita, and N. Sano, 1999. Thermodynamic properties of Si–Al, –Ca,– Mg binary and Si–Ca–Al, –Ti, –Fe ternary alloys, *Mater. Trans. JIM*, 40, 1108–1116.
13. T. Miki, K. Morita, and N. Sano, 1997. Thermodynamic properties of titanium and iron in molten silicon, *Metall. Mater. Trans. B*, 28B, 861–867.

14. R. Noguchi, K. Suzuki, F. Tsukihashi, and N. Sano, 1994. Thermodynamics of boron in a silicon melt, *Metall. Mater. Trans. B*, 25B, 903–907.
15. L. A. V. Teixeira and K. Morita, 2009. Removal of boron from molten silicon using CaO–SiO₂ based slags, *ISIJ Int.*, 49, 783–787.
16. T. Miki, 1999. Thermodynamic property of impurity elements in molten silicon, Doctoral Thesis, The University of Tokyo, Tokyo, Japan.
17. T. Miki, K. Morita, and M. Yamawaki, 1999. Measurements of thermodynamic properties of iron in molten silicon by Knudsen effusion method, *J. Mass Spectrom. Soc. Jpn.*, 47, 72–75.
18. T. Sakata, T. Miki, and K. Morita, 2002. Removal of iron and titanium in polycrystalline silicon by acid leaching, *J. Jpn. Inst. Met.*, 66, 459–465.
19. K. Morita and T. Miki, 2003. Thermodynamics on solar-grade-silicon refining, *Intermetallics*, 11, 1111–1117.
20. C. Wagner, 1962. *Thermodynamics of Alloys*, Addison-Wesley, Reading, Massachusetts, pp. 51–53.
21. G. Inoue, T. Yoshikawa, and K. Morita, 2003. Effect of calcium on thermodynamic properties of boron in molten silicon. *High Temp. Mater. Process.*, 22, 221–226.
22. T. Shimpo, T. Yoshikawa, and K. Morita, 2004. Thermodynamic study of the effect of calcium on removal of phosphorus from silicon by acid leaching treatment. *Metall. Mater. Trans. B*, 35B, 277–284.
23. F. A. Trumbore, 1960. Solid solubilities of impurity elements in germanium and silicon, *Bell Syst. Tech. J.*, 39, 206–233.
24. T. Yoshikawa and K. Morita, 2003. Solid solubilities and thermodynamic properties of aluminum in solid silicon, *J. Electrochem. Soc.*, 150, G465–G468.
25. T. Yoshikawa and K. Morita, 2003. Removal of phosphorus by the solidification refining with Si–Al melts, *Sci. Technol. Adv. Mater.*, 4, 531–537.
26. T. Yoshikawa and K. Morita, 2005. Thermodynamics on the solidification refining of Si with Si–Al melts, in *Proceedings of the EPD Congress 2005, TMS Annual Meeting*, San Francisco, California, pp. 549–558.
27. T. Yoshikawa and K. Morita, 2005. Refining of Si by the solidification of Si–Al melt with electromagnetic force, *ISIJ Int.*, 45, 967–971.
28. T. Yoshikawa and K. Morita, 2005. Removal of B from Si by the solidification refining with Si–Al melts, *Metall. Mater. Trans.*, 36B, 731–736.
29. T. Yoshikawa, K. Arimura, and K. Morita, 2005. B removal by Ti addition in the solidification refining of Si with Si–Al melts, *Metall. Mater. Trans.*, 36B, 837–842.
30. T. Yoshikawa and K. Morita, 2007. Continuous solidification of Si from Si–Al melt under the induction heating, *ISIJ Int.*, 47, 582–584.
31. Y. Nishi, Y. Kang, and K. Morita, 2010. Control of Si crystal growth during solidification of Si–Al melt, *Mater. Trans.*, 51, 1227–1230.
32. K. Morita and T. Yoshikawa, 2011. Thermodynamic evaluation of new metallurgical refining processes for SOG-silicon production, *Trans. Nonferr. Met. Soc. China*, 21, 685–690.
33. T. Yoshikawa, K. Morita, S. Kawanishi, and T. Tanaka, 2010. Thermodynamics of impurity elements in solid silicon, *J. Alloys Compd.*, 490, 31–41.
34. X. Ma, T. Yoshikawa, and K. Morita, 2013. Removal of boron from silicon–tin solvent by slag treatment, *Metall. Mater. Trans. B*, 44B, 528–533.
35. X. Ma, T. Yoshikawa, and K. Morita, 2012. Phase relations and thermodynamic property of boron in the silicon–tin melt at 1673 K, *J. Alloys Compd.*, 529, 12–16.
36. X. Ma, T. Yoshikawa, and K. Morita, 2014. Purification of metallurgical grade Si combining Si–Sn solvent refining with slag treatment. *Sep. Purif. Technol.*, 125, 264–268.
37. H. Morito, T. Karahashi, M. Uchikoshi, M. Isshiki, and H. Yamane, 2012. Low-temperature purification of silicon by dissolution and solution growth in sodium solvent, *Silicon*, 4, 121–125.

38. D. W. Lyon, C. M. Olson, and E. D. Lewis, 1949. Preparation of hyper-pure silicon, *J. Electrochem. Soc.*, 96, 359–363.
39. E. R. Johnston and J. A. Amick, 1954. Formation of single crystal silicon fibers, *J. Appl. Phys.*, 25, 1204–1205.
40. K. Yasuda and T. H. Okabe, 2010. Solar-grade silicon production by metallothermic reduction, *JOM*, 62, 94–101.
41. S. Pizzini, 2015. *Physical Chemistry of Semiconductor Materials and Processes*, John Wiley & Sons, Chichester, UK.

Index

Printed and bound by CPI Group (UK) Ltd, Croydon, CR0 4YY

01/11/2024

01782617-0005